Umberto de Angelis is full professor of physics (retired) at the University Federico II, Napoli, Italy, where between 1971 and 2014, he has taught courses in General Physics, Astrophysics, Plasma Physics and Physics for Astrophysics.

His research activity has been in the fields of Astrophysics and Plasma Physics, mostly in the framework of research grants from the European Community, in the UK (Oxford University, Imperial College London), Germany (Max Planck Institute for Extra-terrestrial Physics, Munich) and Sweden (Royal Institute of Technology, Stockholm).

He has published more than 120 scientific papers in peer reviewed journals, several books of physics for students and four books in popular science (in Italian):

A due passi da noi, Esplorazioni spaziali diTerra e dintorni, Bibliopolis, 1992

Il mondo dell'improbabile homo sapiens, Europa edizioni, 2019

Il problema energetico. Perchè l'italia non può fare a meno del nucleare, GEDI, 2022

Gli esopianeti. La ricerca di vita extraterrestre, Albatros, 2023

From 2010 to 2015, he has been on the Scientific Board of the European Space Agency for the microgravity experiments on the International Space Station.

To my grandchildren, Alexandra, Daniele and Francesca.

Umberto de Angelis

THE EXOPLANETS

Is Homo Sapiens Alone in a
Universe Teeming with Life?

AUSTIN MACAULEY PUBLISHERS™

LONDON • CAMBRIDGE • NEW YORK • SHARJAH

ISBN 9781035828777 (Paperback)
ISBN 9781035828784 (ePub e-book)

www.austinmacauley.com

First Published 2024
Austin Macauley Publishers Ltd®
1 Canada Square
Canary Wharf
London
E14 5AA

20240801

This book is intended for readers without any particular scientific background, I am very grateful to my daughters, Livia and Francesca, and to my wife, Lorella, for being my first readers and for their helpful comments.

Table of Contents

Preface

Where is everybody? Physicist Enrico Fermi posed it back in 1950 during lunch with fellow physicists, igniting decades of debate. Even at a leisurely, slower-than-light pace, the reasoning went, our galaxy could be easily crossed by a spacefaring civilisation within a few million years. The Milky Way Galaxy is presently pushing 14 billion. And while it took four-billion-plus years for technological intelligence to develop on our planet, the galaxy contains plenty of planetary systems of comparable age, as well as others much older.

I propose a possible answer to Fermi's question: "they are all out there, on millions of planets teeming with life, but bacteria do not build spaceships," arguing that although the probability that life exists on other planets in the universe is so close to one that it can be considered an acceptable 'scientific' assumption, the chain of improbable events out of which life evolved on our planet makes the probability of the presence of 'similar' forms of life elsewhere in the universe so close to zero that we can consider it null.

The book follows the long path of man, from the origin of life on earth to the 'improbable' appearance of homo sapiens, after a sequence of highly improbable events, from the structure of the universe with its present mysteries and the search for life in the solar system to the discovery of the exoplanets, where life could be present, to conclude that homo sapiens is probably alone in a universe teeming with life.

1. Introduction: From Crystalline Spheres to Exoplanets

After climbing down from the trees, we understood the nature of fire and lightening, moved the Gods out of Mount Olympus, substituted a star to Apollon carrying the sun around a flat earth on his cart, no longer made human sacrifices to the Gods, understood that the stars do not rotate around us embedded in crystalline spheres moved by angels.

Apollon Drags the Sun

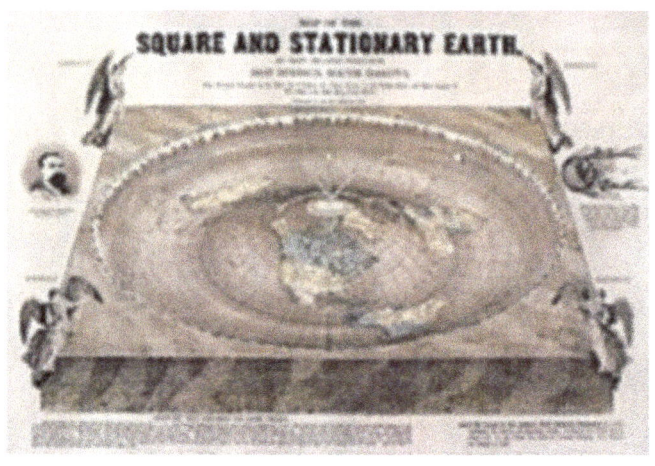

Flat Earth Map (1893). The map contains several references to biblical passages.

We understood that the earth is not flat, and the works of Copernicus and Galileo changed our geo-centric view of the world, an immobile earth at the centre of a closed universe, to the helio-centric view.

We discovered that the universe is not limited by the sphere of the fixed stars (those we can see with a naked eye) but, as Giordano Bruno guessed, it contains innumerable stars grouped together in structures named galaxies. Bruno was probably the first person to grasp that 'stars are other suns with their own planets' and that other worlds, like earth, 'contain animals and inhabitants'.

The whole Aristotelian-Ptolemaic-Theological system of the world became soon incompatible with the discoveries made by Galileo observing the sky with his telescope in 1609, mostly described in his book Sidereus Nuncius.

The first full-view photograph of the planet was taken by Apollo 17 astronauts enroute to the Moon in 1972

The crystalline spheres had already been 'shattered' before Galileo's discoveries. The Danish astronomer Tycho Brahe observed a comet appeared in 1577 and proved that it moves around the sun in an orbit external to the orbit of Venus, that is it 'must have crossed the crystalline spheres'. This left two alternatives: either the spheres did not exist, or the comet had destroyed them!

In his *Sidereus Nuncius*, published in 1610, Galileo reported what he had 'observed' pointing his new telescope to the sky and changed the world forever: the closed and limited world, with the distinction between the earth and the heavens moved by angels, is dead.

The English astronomer Thomas Digges drew a map of the universe where, for the first time, it was not limited by the sphere of the fixed stars, but the stars extended to infinity.

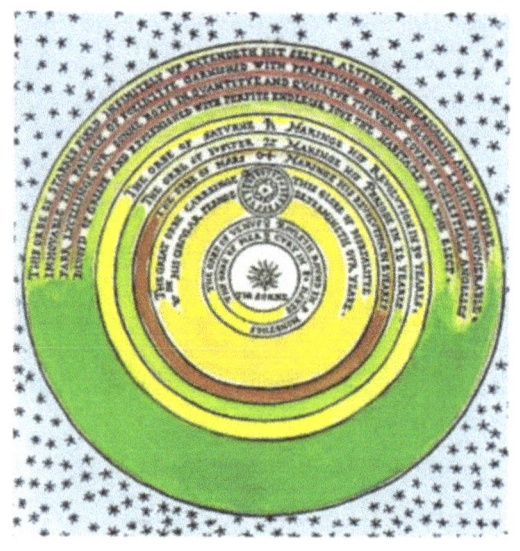

Map of the Universe by Thomas Digges.

New attention is given to the words of Bruno: *[The whole universe] then is one, the heaven, the immensity of embosoming space, the universal envelope, the ethereal region through which the whole hath course and motion. Innumerable celestial bodies, stars, globes, suns and earths may be sensibly perceived therein by us and an infinite number of them may be inferred by our own reason.* [Giordano Bruno, *De l'infinito universo et mondi* (*On the Infinite, Universe and Worlds*),1584].

The first doubts arise about man being alone in the universe:

...Add to this fact the unbearable pride of man who convinced himself that Nature is there but for him, as it be likely that the Sun...had been switched on just to ripen his loquats and grow his cabbage [Cyrano de Bergerac, *Histoire comique des etats et empires de la Lune* (*Comical History of the States and Empires of the Moon*), 1656],

I still find it odd that the Earth should be inhabited, and other planets should not...can you believe that after the Earth has been thus made to abound with life, the rest of the planets have not a living creature on them? Our Sun enlightens planets; why should not every fixed star likewise enlighten planets? [Fontenelle, *Entretiens sur la pluralitè des mondes* (*Conversations on the plurality of worlds*), 1686]

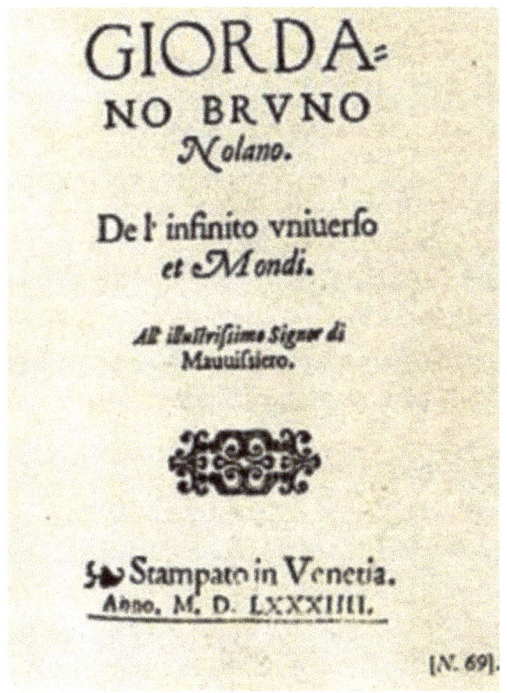

We know how a star is born, its structure and why it shines for billions of years.

We know *today* that the universe was 'born' about 15 billion years ago. The theory describing this 'birth', the Big Bang, seems *today* scientifically more acceptable than other hypotheses, but it leaves many problems still open. And 'last but not least', the universe seems to be full of 'dark matter' and 'dark energy' but we do not yet know what these are.

We know that the sun is one of the billions of stars in one (the Milky Way) of the billions of galaxies in the universe; we know when it was born and when it will die.

We know how the earth evolved from its origin with the solar system about 4.5 billion years ago.

We know the processes from the elements present on the primordial earth led to the appearance of complex organic molecules the basic constituents of the proteins: we have reproduced those processes in the laboratory.

When these molecules interacted to form more complex structures with the extraordinary property to produce copies of themselves—the replicants—it was the first spark of life. The simplest and most ancient forms of life found so far, single-celled prokaryotic organisms (cells without nucleus), are dated up to about four billion years ago, it looks therefore that life, at least in its simplest form, can appear 'almost immediately' where appropriate conditions exist. We do not know yet how the following very improbable steps took place, first from the prokaryotic to the eukaryotic cell (cells with a nucleus) and then to multicellular organisms, but we know that there has been an exceedingly long time to do it: the oldest single-celled eukaryotic organisms are dated about two billion years ago and the first multicellular organisms, the blue algae, about 1.5 billion years ago, therefore almost two billion years passed from the origin to the more complex forms of life. The process leading to the eukaryotic cell may have a very low probability to occur but there has been plenty of time for it to be realised out of all possible combinations.

The earliest known life forms on earth are putative fossilised microorganisms, found in hydrothermal vent precipitates, that may have lived as early as 4.28 billion years ago, relatively soon after the oceans formed 4.41 billion years ago, and not long after the formation of the earth 4.54 billion years ago.

Darwin's theory of evolution has then explained how, from the first living forms, the different species emerged and why complex organs, like the human eye or the bat's ear, do not require the action of a 'watchmaker' building them for an end following a design, natural selection, contrary to selective breeding, acts as a 'blind watchmaker', the title of one of Dawkins' books.

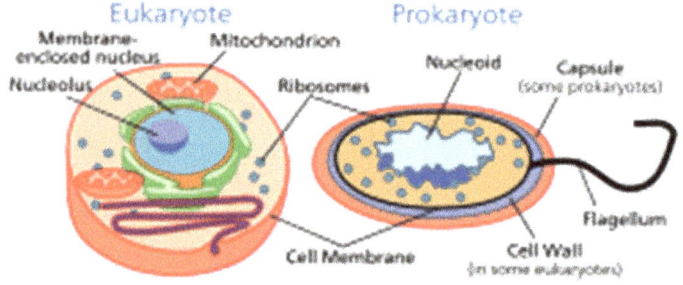

Eukaryotic Cell (left) and Prokaryotic Cell (right)

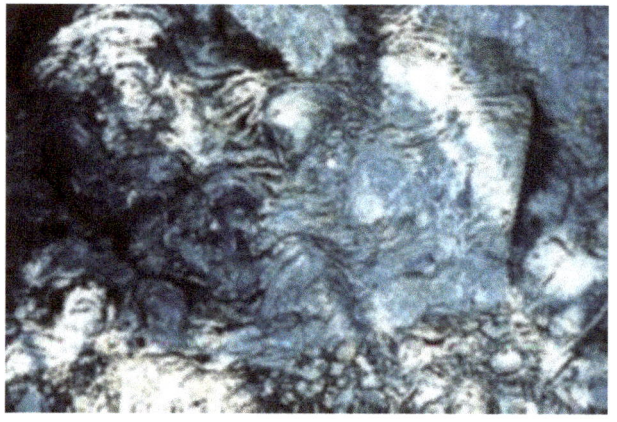

Stromatolites are left behind by cyanobacteria, also called blue-green algae. They are the oldest known fossils of life on earth. This one-billion-year-old fossil is from Glacier National Park in the United States.

Selective breeding transformed teosinte's few fruit cases (left) into modern maize's rows of exposed kernels (right). Darwin's idea was that natural selection can act as a breeder, but without an end or design.

We discovered that, like the sun, other stars have planets, called exoplanets, that is planets external to the solar system. From the number of exoplanets discovered in the last 20 years, we can reasonably assume that, as Bruno wrote, 'innumerable worlds' exist in the universe. We have reasons to believe that the presence of exoplanets in the tiny part of our galaxy explored so far is the rule for the entire universe rather than an exceptional case. If so, there are billions of other worlds and millions of these are likely to be in the 'habitable zone' of their star. We can therefore conclude that the probability that life arose on some of them is so high that we can consider it a certainty. But we also know that life, as we know it, is the result of a long chain of highly improbable events: the probability that these were *all* repeated elsewhere, even given the same initial conditions, is vanishingly small. In conclusion, there is 'almost' certainly life somewhere in the universe, but 'almost' certainly it bears no resemblance to the life forms on earth, except may be for unicellular organisms. For *homo sapiens,* we can drop 'almost'.

2. Darwin's Idea

The Ichneumonidae, a species of wasp, lay their eggs inside other animals, like spiders or caterpillars: they inject their victim with just enough venom to paralyse it without killing it and then pump their eggs into their body. When these hatch, the new born find plenty of... 'Very fresh' food and devour their host from within causing his slow and painful death.

Charles Darwin, who revolutionised our way to understand the world with his theory of evolution, wrote: *I cannot persuade myself that a beneficent and omnipotent God would*

have designedly created the Ichneumonidae with the express intention of their feeding within the living bodies of caterpillars. (Charles Darwin in *Letter to Asa Gray*, 1860)

Charles Darwin (1809–1882)

One of the seven proofs of the existence of God proposed by Thomas Aquinas is the 'teleological argument' (or of the design): everything in the world seems to be designed for a scope, the designer is God.

The works of nature are so perfect that there is no room for perfecting them, wrote the English naturalist John Ray in his essay *The wisdom of God manifested in the works of the creation* (1691).

'Natural Theology', developed between the seventeenth and eighteenth centuries, particularly in England, and culminated with the 'Bible' of the natural theologists, the text published in 1802 by William Paley 'Natural Theology: or Evidence of the Existence and Attributes of the Deity', is a re-visitation, in the framework of the new knowledge of nature, of the old argument of 'the design'. Its objective was to 'demonstrate' God's existence and goodness through the contemplation of nature's marvels and the benevolence it shows to man, the superior creature. The enormous variety and complexity of the living species and the evidence of a 'scope' in the design of all creatures, the perfect adaptation of the structure of each living being to its functions and to its habitat, can only be explained if we admit that all living species have been created 'as we know them' by a good and omnipotent 'designer'.

There is a great abundance of examples, from the wonderful harmony of the Newtonian system of gravitation to the perfect functionality of organs such as vertebrae, eye and gall bladder, from the careful location of each species in its habitat (if left to chance, the distribution of species could have led to absurd results, like polar bears in the equatorial jungle or kangaroos on the highest mountains) to the rattles in the rattle-snake, to warn potential victims of the peril.

All this requires intelligence and scope, a design, that is a Creator.

The argument of the 'design' is still alive today in the debate between creationism and evolutionism.

Creationists assume that living species have been created by God *as they are*, only a deity, a perfect 'watchmaker', could realise structures so complex, so well adapted to their scope, more perfect than a Swiss clock like, for instance, the human eye.

Leaving aside that, as we know now, the human eye is far from being perfect, and if it was created by an omnipotent God, he did a lousy job; it is now scientifically confirmed that the species have evolved. And it is also scientifically confirmed that they evolved according to Darwin's original idea.

Here is the reply of the naturalist David Attenborough, well known for his splendid documentaries about nature, in an interview about creationism:

My response is that when creationists talk about God creating every individual species as a separate act, they always instance hummingbirds, or orchids, sunflowers and beautiful things. But I tend to think instead of a parasitic worm that is boring through the eye of a boy sitting on the bank of a river in West Africa, [a worm] that's going to make him blind. And [I ask them], "Are you telling me that the God you believe in, who you also say is an all-merciful God, who cares for each one of us individually, are you saying that God created this worm that can live in no other way than in an innocent child's eyeball? Because that doesn't seem to me to coincide with a God who's full of mercy..."

The essence of the Darwinian revolution is in that it has shown that there is an alternative to the 'design': the evolution of species through the accumulation—over long periods of time—of occasional mutations useful in the struggle to survive and therefore naturally 'selected' for the preservation and transmission to the offspring. In the words of Darwin:

Owing to this struggle for life, any variation, however slight and from whatever cause proceeding, if it be in any degree profitable to an individual of any species, in its infinitely complex relations to other organic beings and to external nature, will tend to the preservation of that individual, and will generally be inherited by its offspring. [Darwin *The Origin of Species*]

The idea of natural selection had also been proposed by the naturalist Alfred Russel Wallace in a work published shortly before *The Origin of Species*, but Darwin was probably not aware of it before both works were presented together at the Linnean Society in 1859 and only in Darwin's work, we find an accurate, detailed and scientifically valid justification of the idea.

The fundamental points at the basis of *The Origin of Species* can be summarised as follows:

Travels and deep knowledge of nature

In the first half of the nineteenth century, England dominated the seas all over the world and the travels of naturalists and geologists aboard English ships had provided new and deep knowledge about the flora and fauna of all parts of the world.

The young Darwin had the good luck to be taken as a naturalist on board the ship Beagle for a 40.000 miles voyage, visiting places from South America to the islands of the Pacific and Atlantic Oceans, from Australia to South Africa.

In this voyage, that Darwin himself acknowledged as the origin of his career, the long excursions on land allowed him to observe, collect and interpret flora, fauna and geological formations of completely different habitats, from the Brazilian jungles to the mountain peaks in the Andes, thus reaching a knowledge, unique at his time, of the geographical distribution of species, the relationship between extinct and living species, the particular structure, advantages or disadvantages of some organs and of rudimentary organs like the eye of the blind *tucutucu* (see Charles Darwin *A naturalist voyage or The voyage of the Beagle)*.

All the observations made in the course of this voyage, systematically reported in his diary, allowed Darwin, after years of 'meditation', to insert in *The Origin of Species* detailed examples and arguments in support of his idea about the evolution of species.

But already during his voyage, some of the things he observed undermined his belief in the theory of immutable species, two in particular-the great similarity of structure between extinct (fossil) and living species seems to suggest descendance from common ancestors, that is evolution and the extraordinary likeness of finches in the different Galapagos islands, clearly the same species but with slight variations from island to island, in the beak and colour, for instance, such as to make each finch particularly adapted to the habitat of the island on which it lives, seems to suggest a common ancestor (which lived perhaps when the islands were not yet separated) from which, in diverse habitats, isolated after the separation of the islands, 'different forms' evolved in response to the diverse environments. Once again: species can change!

Darwin's Finches

Darwin himself wrote that it took several years, after his return from the voyage, to let these ideas 'ripen', convincing him that species are not immutable but evolve.

The evolution of species was not a new idea in Darwin's time, evolutionist views could be found in some philosophers of ancient Greece and, in more recent times, in philosophers and naturalists as Diderot, Buffon, Lamarck and even Darwin's grandfather, Erasmus Darwin.

There were two basic reasons why these had never been seriously considered, even by Darwin himself:

1. The age of the earth, still believed to be a few thousand years, was not sufficient for the 'long' work of evolution.
2. No one had provided an acceptable explanation of 'why' species evolve.

Lamarck (1809) for instance had suggested the influence of the habitat as a possible cause, the 'need' of an organism to change for a better adaptation to its environment and, once even a slight change has occurred in some organ due to the 'effort' of an individual to better its performance, it is passed on to its offspring... and so on. Examples of Lamarckism, that is adaptation to the environment, include the giraffe developing its long neck in the continuous effort to reach higher leaves on trees, the sole developing a flat body to swim in shallow water and moving both eyes to one side to swim on the other side, the snake moving its eyes on top of the head to see better any approaching danger while crawling and developing a long, sensitive tongue to identify preys and danger up front.

But Lamarck's theory has no explanation for the 'mechanism' through which the influence of the environment can modify an organ until a better adaptation is reached.

To have any credit evolutionary theories needed to find a 'plausible' mechanism to explain their action and to have sufficient geological time to operate.

For the first requirement, Darwin's extremely clever idea was to consider the accidental mutations as the starting basis of the evolutionary process.

The second requirement was provided to Darwin by the:

New discoveries of geology: fossils and the Earth's age

Near the end of 1600, two scientists, the English R Hooke and the Danish N Stensen (Latin Stenone) were the first to assert that fossils are not 'nature's jokes', as they were mostly considered, but that their presence in regular strata on the earth is the proof that such strata form a temporal sequence, having been deposited one on top of the other starting from the lowest one.

In 1700, scholars still believed that the earth had been created a few thousand years before, but already at the beginning of 1800, many were convinced that it must be much older, as suggested by the superposition of strata in the earth's crust; once it was understood that such stratification represents a temporal sequence. These strata form a roughly vertical structure, with the oldest at the basis and the most recent ones near the earth's surface. By the end of 1700, it was clear that these strata are a direct representation of the history of the earth: each layer had once been on the surface, hence from the 'maps' of the successive exposition of the layers on the surface it was possible to deduce their historical sequence, that is the history of the earth.

The great French chemist Antoine-Laurent Lavoisier (1743–1794) was the first to suggest, including sketches of the vertical sequence of sediments in the geological maps: this allowed to add a vertical dimension to the two-dimensional vision of the geological maps and hence the reconstruction of the geological history of a region from the variations in the underlying strata from each zone to the next.

Lavoisier became interested in the new science of geology only at a late time, after he had given, as a young man, fundamental contributions to the development of chemistry, and unfortunately his career as a geologist, started during the French Revolution of 1789, was literally 'cut' in 1794 when he lost his head on the guillotine.

He had only time to write a single paper on his research, published at the very beginning of the Revolution (1789), with the explicative title:

General observations on the recent horizontal beds that have been deposited by the sea, and on the consequences that one can infer, from their arrangement, about the antiquity of the earth.

The most important contribution by Lavoisier was the understanding and use of the difference between the strata deposited on the seabed (pelagic) and those deposited on the coasts. The first, containing the remains of shells and other maritime animals, must have been formed because of a slow accumulation and therefore in a very long time; the latter originate from the erosion of coastal rocks due to the action of the sea. The way these two sediments alternate in the vertical structure of the strata is therefore a clear indication of the ascending and descending cycles of the sea level and given the very long times required for the formation of pelagic strata, the presence of numerous strata reveals the great antiquity of the earth. (On the work of Lavoisier as geologist, see Gould *The Lying Stones of Marrakech).*

The effects of the oscillations of the sea level have also been important in the development of a new geological theory, uniformitarianism, which replaced in the course of 1800 the theory of catastrophism.

According to the latter, the present state of the earth had been formed in a series of natural 'catastrophes', like volcanic eruptions, earthquakes, and floods, in a 'short' time following its birth as a mass of fused material emitted from the sun, as suggested by the most accepted theory on the birth of the earth. With the successive cooling of the earth, the frequency and intensity of the catastrophic events progressively decreased leaving room to a much quieter planet.

The 'uniformitarians' assumed a radically opposed point of view: the earth's geology has to be studied with the hypothesis that the present state is the result of the same processes as we observe today (volcanic activity, earthquakes, erosion…), with limited frequency and intensity but acting over very long times. Apparently, 'catastrophic' events are only the result of the accumulation, in the infinity of geological time, of small changes, like the excavation of deep canyons particle after particle. It is a uniform history of the evolution of our planet which, consequently, must be much older than previously thought.

As suggested by SJ Gould (in *The Lying Stones of Marrakech),* it was during a visit to my city, Naples, that two of the top proponents of the new geology, Charles Lyell and Charles Babbage, found a particular support to the uniformitarian idea.

On the frontispiece of the famous Lyell's text, *Principles of Geology* (which Darwin had with him in his travel around the world), there is the figure of the three columns of the 'temple of Serapis' in the town of Pozzuoli, near Naples: from the level of the holes pierced in the marble columns by a species of marine bivalve, Lyell deduced the proof of the slow and gradual oscillations of the sea level in the last two millennia, contrary to the catastrophic image of Mount Vesuvius.

The Three Pillars of Pozzuoli

The new science of geology therefore allowed Darwin all the time needed for the slow action of evolution. Also important were the developments of palaeontology and anatomy, mainly due to the works of Cuvier in France and Owen in England, which started to provide the correct interpretation of fossils.

Fossils had been known for a long time, already in the Natural History of Pliny the old, who died in the eruption of Mount Vesuvius which destroyed Pompeii, we find descriptions of fossil shells on mountain tops, correctly interpreted by Pliny as evidence of the raising of submerged areas.

We know today that fossils are the remnants of ancient organic beings, and as such, they are fundamental in the study of the evolution of species.

In the seventeenth century, the dominant interpretation of the nature of fossils was that they were of inorganic material, parts of rocks 'modelled' by 'natural forces' to resemble organic beings, a 'symbolic correspondence' due to a 'resonance' between natural forces acting in the inorganic world and 'vital forces' acting in the animal world.

Such 'resonance' between the inorganic and animal worlds was brought to extreme consequences by a sort of 'medical theory' according to which the 'resonant forces' between minerals and living beings, the resemblance of some rocks to human organs, could be used to cure ailing organs. A stone resembling a foot, pulverised, and ingested, was suggested as a cure for gout, while a stone resembling some genital organ could help to cure sexual disorders.

Like the famous 'vulva stone', later recognised as the fossil of a brachiopod, whose structure resembled the female organ on one side and the male organ on the other: in a text of 1665 on the properties of such stones we read:

And I think it not silly to believe, especially given the form of these objects...that, if worn suspended around the neck, they will give strength to people experiencing problems with virility...(cited by Gould, *The Lying Stones of Marrakech*).

The idiocy of such arguments can be compared to the ointment recommended in medieval times to cure wounds from guns by spreading it not on the wound but...on the gun!

When finally, by the end of 1700, the evidence of the organic nature of fossils became overwhelming, creationists were forced to change their interpretation: it is true that fossils are the remains of extinct species, but they have all been created in successive times. When a species was no longer perfectly adapted to the changing environment, the good Lord, faithful to the 'intelligent design', would drive the old species to extinction, substituting it with a new one, perfectly adapted to the new habitat.

Even today, creationists hold a similar interpretation of fossils as a 'continuous creation', finding support from some cases of discoveries where there is an apparent 'sudden' substitution of one species with another without intermediate forms. We shall discuss this point later.

By the beginning of 1800, the nature of fossils as remains of extinct species was fully accepted and the similarity of structure and functions between fossils and living species was starting to be appreciated even though, as Darwin himself acknowledged, the geological record was still too poor and fragmented to provide proofs of the gradual evolution of extinct into living species.

An obvious proof would have been the presence of fossils of 'intermediate species', or 'varieties' which, during the very long and slow evolution, should be present in great amount in fossil deposits but were not found.

Darwin is well aware of this particular difficulty with his theory and in a chapter of his book, entirely dedicated to this topic, he shows, with many examples and brilliant arguments, that the few cases known at his time are more compatible with his theory than with continuous creation and that the absence of great amounts of intermediate fossils can be ascribed to the particular modes of sedimentation of the strata where such fossils should be found, as well as to the poverty of the geological record:

Only a small portion of the surface of the earth has been geologically explored, and no part with sufficient care, as the important discoveries made every year in Europe prove... [Darwin *The Origin of Species*]

And now we know many examples of fossils intermediate between species, also for man. There are still cases though where a new species seems to 'appear from nowhere' for the joy of creationists, as mentioned before. Such cases led palaeontologists, Eldredge and Gould in particular, to propose, in the framework of natural selection, a new evolutionary mechanism based on periods of 'fast' evolution followed by 'static' periods where a species maintains its equilibrium 'resisting' to the forces of natural selection. But, as shown by Dawkins (*The Blind Watchmaker*), these 'holes' in the fossil records are perfectly consistent with Darwin's theory, they are explained by the same Darwinian process of formation of species, which, to avoid mixed procreation, requires the geographical separation of the initial species from the one undergoing modifications. The new species may eventually migrate back to the original area, and if better adapted to the habitat than the old species, may cause its extinction and replace it. In the fossils of this area, the new species will seem to appear suddenly simply because its evolution took place in a different area and the intermediate forms are to be found elsewhere!

The *Origin of Species* has its roots in the extension to the natural world of two ideas: the natural selection in the struggle for life from the theories on population of Malthus and Spencer and the exploitation, by breeders, of the occurrence of occasional mutations in domestic species to produce modified species through the 'artificial' selection of the desired characteristics.

The theories of Malthus and Spencer extended to the natural world.

In his *Essay on Population* (1798), Thomas Malthus stated that, since the population grows at a faster rate than available resources, especially food, without preventive checks (like delaying the age for marriage) or external conditions (like wars, famine, epidemics), the production of resources will not sustain the population growth, which will always tend to outrun food supply. Malthus was opposed to England's welfare policy for the poor, which could only increase the demographic development, leading to a ruinous over-population.

Malthus' theory was given an 'evolutionary' interpretation (only for the human species) by the sociologist Herbert Spencer. In his 'Theory of Population' (1852), Spencer assumes that the lack of equilibrium between population and resources is the cause of the 'struggle of men for survival', with the elimination of the 'unfit' and the conservation of the better 'fit'.

The 'survival of the fittest (or strongest)' often wrongly considered as the essence of Darwin's theory, was used by Spencer, never by Darwin.

Darwin knows that the struggle to survive is not exclusively human but is also present in the animal and vegetable worlds, where the enormous growth rate would lead to a population explosion if it were not checked by predators or changes in the habitat or climate, that is by all the complex interactions of every living being with nature and other living beings, interactions leading to the extinction of the 'weakest', the least fit to their environment, and to the survival of those with characteristics for a better adaptation. With his words:

This is the doctrine of Malthus, applied to the whole animal and vegetable kingdoms. As many more individuals of each species are born than can possibly survive; and as, consequently, there is a frequently recurring struggle for existence, it follows that any being, if it vary however slightly in any manner profitable to itself, under the complex and sometimes varying conditions of life, will have a better chance of surviving, and thus be 'naturally selected'. From the strong principle of inheritance, any selected variety will tend to propagate its new and modified form. [Darwin *The Origin of Species. Introduction*]

But what is the 'origin' of these slight variations that allow a better adaptation?

Darwin's clever idea is to look at the mutations: occasionally a newborn has some tract 'different' from his parents and from the general characteristics of his species. Darwin does not know why this happens, nor the laws governing such phenomenon, but he knows, from the selection of domestic animals, that mutations occur.

Artificial selection of domestic species: Darwin's pigeons.

What is surprising when looking at the enormous variety of plants and domestic animals is their adaptation not to their own 'good' but to man's desires. Breeders, gardeners, and farmers have produced animals and plants with 'chosen' characteristics, either for their usefulness to man or for their 'beauty' (still according to man's taste). The differences we observe between varieties of the same species can be extraordinary: think of dogs, all descended from the grey wolf through a long selective action on the part of breeders.

How did they reach breeds with characteristics so different between themselves and from the common ancestor as shown in the figure?

Darwin was a breeder of pigeons, was in touch with the best breeders of his time and a member of two clubs of pigeon lovers. He could therefore observe directly the astonishing differences in breeds like the English carrier, the short-faced tumbler and the common tumbler, the runt, the barb, the pouter, the turbit, the jacobin, the trumpeter and the laugher, the fantail. The differences were not only exterior (beak, skull, eyelids, nostrils, feet, neck, wings and tail, feathers, ways to coo and fly) but also in the skeleton (development of the bones of the face in length, breadth and curvature, the number of sacral and caudal vertebrae and of ribs, the size and shape of the apertures in the sternum) and in the structure (size of crop and oesophagus, oil-gland, number of primary wing and caudal feathers, number of scutella on the toes, skin between the toes). Also, variable was the form and size of the eggs, and manner of flight, voice and disposition can present remarkable differences in some breeds.

And yet Darwin was convinced, and gave the reasons why, that no matter how big the differences, the commonly shared opinion of naturalists was correct: they all descend from the rock-pigeon (Columba Livia).

Darwin noted that pigeon fanciers had created many kinds of pigeon, such as Tumblers (1, 12), Fantails (13), and Pouters (14) by selective breeding.

The same must be true for all the animal and vegetable varieties perfectly adapted to our needs: breeders, gardeners and farmers must have always started from a 'wild' ancestor, an individual with basic structural characteristics similar to those of all the varieties produced. The examples are innumerable. Fruits like the strawberry and the pear, flowers like the dahlia, farm products useful in the diverse seasons, the workhorse and the racing-horse, the greyhound and the dachshund, the fighting and the hen house cocks, obviously did not live in nature as we know them once domesticated, they are man's products. One particular example reported by Darwin is the recent (at his time) variety of strawberries with different size, produced by gardeners in the last decades.

To produce varieties that are useful for him, man takes advantage of the occasional mutations in a plant or animal. It can be observed that seedlings born from the same fruit or the cubs in the same litter sometime show differences between themselves and with parents, during reproduction an occasional, unforeseeable mutation of some character has been produced in the offspring. It can also be observed that, once an individual begins to change due to effect of a mutation, the change persists and is transmitted to its descendants.

Breeders, gardeners, and farmers have made use of such observations, planting the seeds of the seedling (or using the cub to breed) with the desired characters and repeating the procedure every time a new variation in the right direction appeared in the offspring, till the desired result was achieved.

This 'artificial' selection, which has produced animals and plants adapted, in their structure and/or behaviour, to desires and needs of man, is therefore the result of the occasional mutations in the process of reproduction and of man's ability to select and accumulate the useful mutations in the chosen direction.

A tulip flower exhibiting a partially yellow petal because of
a mutation in its genes.

The road towards the theory of 'natural' selection can now proceed in three steps:

-Since mutations are related to the reproductive system, we are allowed to assume, even as we do not know their causes and laws, that they have always occurred in the wild as in the domesticated state.

-In nature, we find differences between individuals not only of the same species but also belonging to varieties intermediate between species. Such differences are often so slight and gradual that naturalists do not always agree on the classification of a variety as such or as a new species. From his observations in the Galapagos, Darwin had derived a clear understanding of the ambiguity that can occur in the decision between variety and species, and in The *Origin of Species*, he gives many examples of such ambiguity, discussing the characteristics of varieties and species in the natural world, based not only on his own knowledge but on the works of zoologists and botanists as well, reaching the conclusion that varieties represent 'evolutionary steps' in the transition between species through the accumulation of structural differences in a given direction. (*The Origin of Species*, chapter II).

The high growth rate of all living beings leads necessarily to a 'struggle for survival', intended as the role of the complex dependence of every living being from other living beings and from the environment in determining the survival of an individual and its ability to reproduce. Darwin presents many examples of such struggle to survive in the vegetal and animal kingdoms, showing how complex can be the interactions between organisms determining survival or extinction, for example how the number of flowers in some region depends on the number of...cats in that region. (*The Origin of Species*, chapter III).

The theory of natural selection

So then, if mutations exist and those useful to man are exploited to select, in relatively short times, animals and plants according to his desires, how can one doubt that mutations, useful not to man but to each organism in its struggle for survival, have taken place during innumerable generations?

And, if they happened, how can one doubt that individuals with some advantage on the others, that is with mutations however slightly advantageous for survival, had better chances to survive and procreate while individuals with mutations however slightly disadvantageous, that is harmful in the struggle for survival, have been eliminated?

The preservation of favourable variations and the rejection of injurious variations, I call Natural Selection [Darwin *The Origin of Species*].

When some useful variation is produced, to what extent is it due to the action of natural selection or to environmental conditions? For instance, is the tick and warm fur of those animals living in cold climates since the individuals who were better clad, due to the accumulation in time of occasional favourable mutations, had always a better chance to survive, or to a direct action of the climate?

Darwin's conclusion is that environmental conditions have an influence not so much on the variability in the structure of organisms, but rather on their reproductive system. Since mutations happen during reproduction, followed by the accumulation through natural selection of all the favourable 'slight' variations until the 'final product' is reached, we can say that environmental conditions are only an 'indirect' cause of variations of organisms.

It is then no wander if the 'final product', achieved after the accumulation in geological times of favourable characteristics, shows a perfect adaptation to the living conditions in the habitat where it evolved. All over the world, natural selection is alert to check any variation, however small, preserving and accumulating those useful to an organ, eliminating those not useful or harmful, in a never-ending task for the 'good' of every living being. Think of the colours of birds and insects, 'designed' by natural selection to provide them with the best camouflage to escape predators. Or the existence of two varieties of wolves in the United States, one faster with longer legs, the other more massive with shorter legs: the first lives where deer are abundant, the second is a predator of sheep. Or the perfect mutual adaptation of some flowers and bees, with structures modified so that both have the maximum advantage from the action of nectar collection.

In Darwin's theory then the varieties of animals and vegetables of a given species represent just initial steps towards a new species, the small differences between organisms of two varieties will tend in time, due to natural selection, to become larger and larger until the two varieties will differ so much that they can be classified as two distinct species, both having structure and habits better adapted to the living conditions of the initial organisms, while the intermediate forms that had variations not useful or harmful for survival will be extinct.

This is the principle of 'divergence of characters', a consequence of natural selection: the modified descendants (varieties) of a given parent will tend, in the struggle for survival, to diversify more and more to breed in new spaces and habitats, the greater the differences, the greater the probability to survive, the small differences between varieties will therefore tend to increase from generation to generation.

The result is represented by Darwin as a tree, 'the tree of life'. The green twigs are the existing species, the branches were once twigs when the tree was younger, some of them do not produce buds anymore, they represent the extinct species; on those which still bud, every bud produces new buds by branching, as varieties produce new varieties and species, with a common 'ancestor': the initial bud. All these common ancestors have in turn a common ancestor: the branch. And all the branches are born from the trunk, born from a seed. If we could go back in time along the tree of life, we would eventually get to know the seed, the progenitor of all living and extinct organic beings.

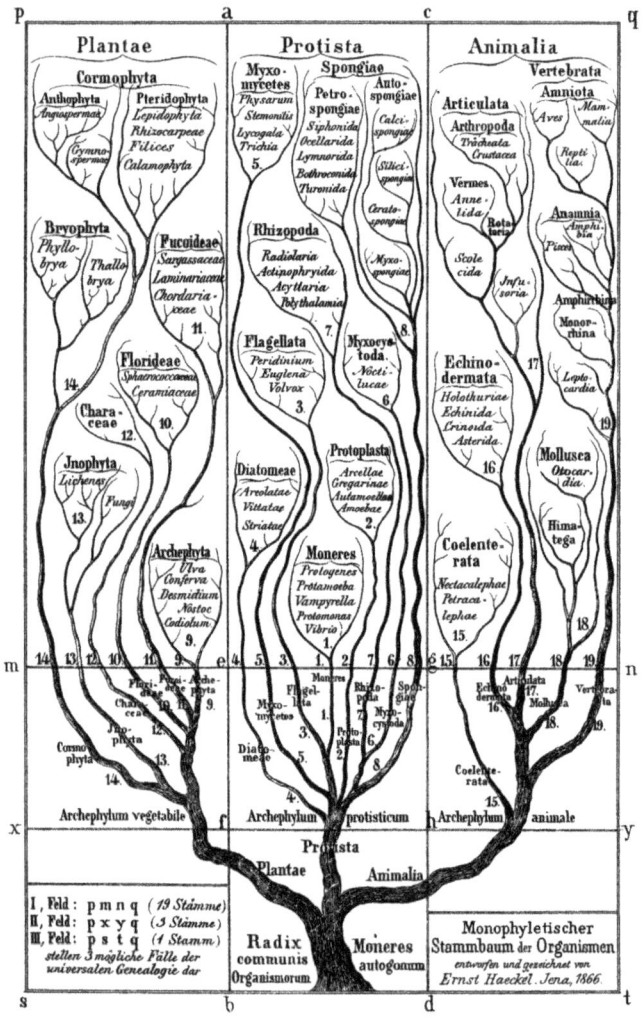

Haeckel's Tree of Life in *Generelle Morphologie der Organismen*
(1866)

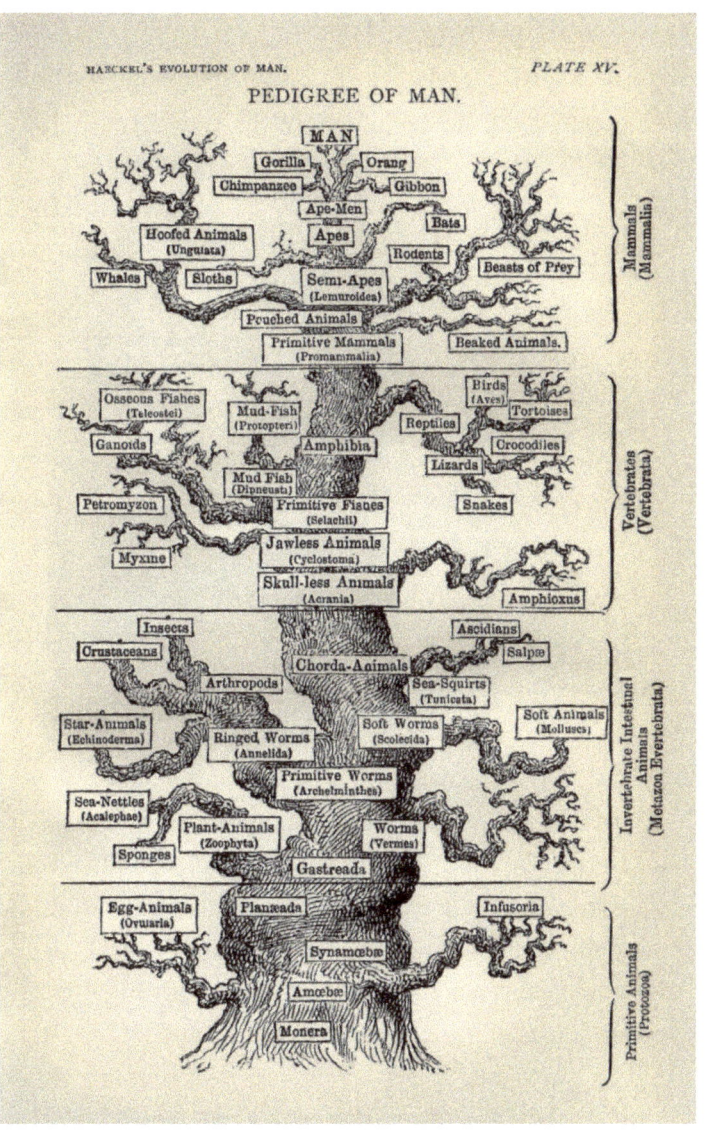

The Tree of Life as seen by Haeckel in *The Evolution of Man*
(1879)

Every species has characteristics quite different from the others, has its own twig, with the only exception of the duck-billed platypus, a small, furry, semi-aquatic Australian animal, which has intermediate characters between mammals and birds: it lays eggs but gives milk to its young.

Platypus in Broken River, Queensland

Darwin's theory of evolution states therefore that natural selection can give origin to species. What natural selection *cannot* do is to produce changes in an individual, useful not to the individual and to its species but to a different species. This would contradict the working mechanism of natural selection (the advantage of the individual) and Darwin writes *...though statements to this effect may be found in works of natural history, I cannot find one case which will bear investigation.*

If Darwin had limited his work to the formulation of the idea of natural selection, The *Origin of Species* would not be the most revolutionary scientific text ever published, as it is considered by many. Its extraordinary success is based on the fact that after introducing the idea of natural selection, Darwin tackles an essential task: to prove that natural selection is the alternative to the 'design' for all the living conditions in the vegetal and animal kingdoms considered by Natural Theology as evidence of the need for a 'designer'.

Natural Selection can act only by the preservation and accumulation of infinitesimally small inherited modifications, each profitable to the preserved being; and as modern geology has almost banished such views as the excavation of a great valley by a single diluvial wave, so will natural selection, if it be a true principle, banish the belief of the continued creation of new organic beings, or of any great and sudden modification in their structure. [Darwin *The Origin of Species*]

The *Origin of Species* is a revolutionary text because it demolishes, point after point, the bases of the text by William Paley *Natural Theology: or Evidence of the Existence and Attributes of the Deity*.

From Paley's introduction: *...suppose I pitched my foot against a stone, and were asked how the stone came to be there, I might possibly answer that...it had lain there forever...But suppose I had found a watch upon the ground...I should hardly think of the answer which I had before given...that the watch had always been there...the watch must have had a maker...*

For Paley, the watchmaker is of course God, Darwin proves that, speaking of a watchmaker, it is a 'blind watchmaker', it does not know the future of its actions, does not plan its consequences, its products have been built step by step, without any 'design' (see Dawkins, *The Blind Watchmaker*).

Paley's work is admirable for the detailed description of the natural world and of the surprising adaptation of animals to their lifestyle. Paley examines two possible alternative explanations for such perfect adaptation, either the formation of an organ for reasons different from the one for which the animal discovers afterwards that it can be used or the formation, step after step, with a slow evolution towards the 'final' state, a process like Lamarck's theory, of which probably Paley was not aware yet.

But both alternatives are criticised and discarded by Paley, for instance in the following pieces as cited by Gould [*Eight little piggies*]:

It is possible to believe that the eye was formed without any regard to vision; that it was the animal itself which found out, that, though formed with no such intention, it would serve to see with?

If it be suggested that this proboscis may have been produced in the long course of generations...I would ask, how was the animal to subsist in the meantime, during the process, until this elongation of snout were completed?

The only alternative left for Paley is the action of a benevolent Creator. Darwin, who admired Paley's work, accepted all the reported facts, but proposed a different interpretation. There is some irony in the fact that in Darwin's interpretation the benevolent God is replaced with the continuous extermination of the 'not fit'.

We cannot follow here the detailed analysis presented by Darwin in the *Origin* of all those cases where natural selection is shown as the alternative to creationism, from the wonderful adaptation of organs to living conditions and the development of instinct, from the geographical distribution of species to hybridism, from the rudimentary organs to embryology, up to the 'war horse' of Natural Theology, the organs of extreme perfection like the eye, whose gradual evolution, may be starting from an occasional mutation by which some nerve became sensitive to light, is compared by Darwin, as also Paley did, to the work of man in perfecting the telescope. But, while Paley's conclusion was that an organ requiring the perfect combination of so many parts could only have been created, Darwin provides two arguments to justify its progressive construction: every modification, however slight, if it results in a better visive perception is helpful in the struggle for survival to the individual and its offspring inheriting it and the time required for the accumulation small successive modifications is extremely long. May be a new mutation resulted in the formation of a transparent membrane on a light-sensitive nerve. The individual with such new characteristic has obviously some advantages with respect to other individuals of his species: he can distinguish day and night and be better aware of the presence of predators. With a long series of small successive mutations, such membrane could separate in layers of different thickness, form and density and natural selection would work to preserve any small variation resulting in the formation of more distinct images which, however slightly distinguished from the previous state, increase somehow the advantages in the struggle for survival. And so on for millions and millions of years, for millions and millions of individuals, with the action of natural selection always present to favour the survival of those with useful variations.

Fossils can tell us nothing about organisms with rudimentary eyes, since soft parts decay fast, but Darwin points out that among living animals there are some species of crustaceans presenting gradual differences in the eye structure and he concludes: *...and may we not believe that a living optical instrument might thus be formed as superior to one of glass, as the works of the Creator are to those of man?* [Darwin *The Origin of Species*].

Dawkins's book *The Blind Watchmaker* contains detailed discussions of the formation, by accumulation of small variations, of complex organs like the eye, the lungs, the wings, the ear and mimicry, that is of the power of natural selection to produce 'optimal designs'.

With Darwin's words: *If it could be demonstrated that any complex organ existed, which could not possibly have been formed by numerous, successive, slight modifications, my theory would absolutely break down. But I can find out no such case.* [Darwin, *The Origin of Species*]

The evolution of man is not discussed in the *Origin* but in a later book *The Origin of Man and Sexual Selection,* published in 1871.

The idea that man, as all other living beings, originated from presently extinct ancestors lower in the geological sequence, had already been suggested in Darwin's time, but without any proof.

No proof could possibly be found in the fossils at that time, the first fossil to be considered as man's ancestor was discovered only in 1891 in the island of Java, the pithecanthropus erectus, with more ape-like than human-like characters.

Darwin assumes that his theory of natural selection be valid also in man's struggle for survival and shows, through a detailed analysis of the morphological, physiological, and psychological affinities of man with other mammals, to what extent man is similar to other mammals and the traces of his descendance from inferior forms.

Darwin's conclusion on man's evolution has been and still is erroneously interpreted as the 'descendance of man from apes', from the famous question asked by Wilberforce during a debate between evolutionists and creationists *if Dr Huxley descends from an ape on the part of father or mother?*, to misleading iconographies (unfortunately still found today) on the 'path of man', as the one shown in the following figure.

We need to remember that on Darwin's tree of life man and ape are on different branches, even though they share a common ancestor.

A more serious objection to Darwin's theory, by F Jenkin, was based on the hereditary mechanism. In Darwin's time, it was believed that newborn inherited a mixture of the parents' characters and, if so, any mutation occurred in one of the parents would have been 'diluted' in the coupling with the other parent, the 'part' of the mutation inherited by the descendant 'diluted' again when it would couple in its turn…and so on. Therefore, any variation in an organism would rapidly disappear instead of accumulating generation after generation, which clearly undermined the mechanism of natural selection.

The problem had already been solved by the Austro-Hungarian monk Gregor Mendel, who had demonstrated that characters are transmitted integer from generation to generation, not as a mixture of both parents' characters. For years Mendel had performed experiments with peas of different varieties (yellow or green colour, smooth or rough skin), observing the characters of the peas born either from self-pollination of the same variety or from the crossing of different varieties. From his observations, he could show that plants with only one of the characters (for instance yellow peas) were born in a first generation, but also the other character (for instance green peas) was reproduced in the second generation with a constant ratio of 3:1. In other words, the hereditary characters are transmitted as separate units maintaining their individuality and are preserved from generation to generation. Mendel called 'Elemente' these 'hereditary units', today we call them 'genes'.

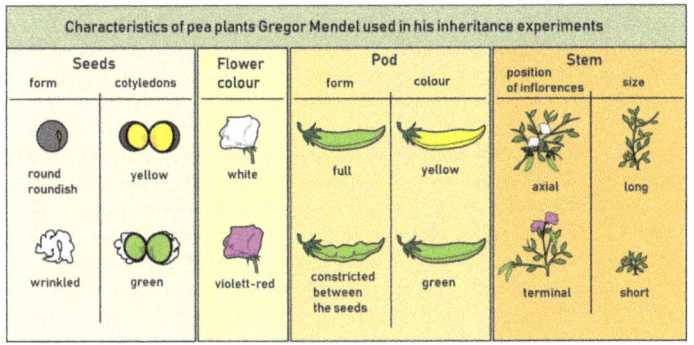

Seeds		Flower	Pod		Stem	
form	cotyledons	colour	form	colour	position of inflorences	size
round roundish	yellow	white	full	yellow	axial	long
wrinkled	green	violett-red	constricted between the seeds	green	terminal	short

Characteristics of pea plants Gregor Mendel used in his inheritance experiments

Characteristics Mendel used in his experiments.

From his results, Mendel could derive two laws, which are still today at the basis of the science of genetics, but Darwin did not know of Mendel's work, which in fact was ignored till the beginning of 1900, when its importance was finally understood.

Darwin was obviously conscious of such difficulty with his theory, he accepted that the mixing of characters would certainly put a limit to the survival of 'big variations', but he observed that the mechanism of natural selection is based on the accumulation of imperceptible variations which are not deleted by the mixing of characters (see Gould's essay on F Jenkin in *Bully for Brontosaurus*).

Darwin does not know the causes or the laws governing mutations, he knows that these happen and assumes that they are transmitted unaltered to the offspring.

We know today that the geological time available to evolution is more than two and one half billion years.

Many fossils of intermediate species have been found, missing steps in the long succession of species, including those revealing the evolutionary history of man.

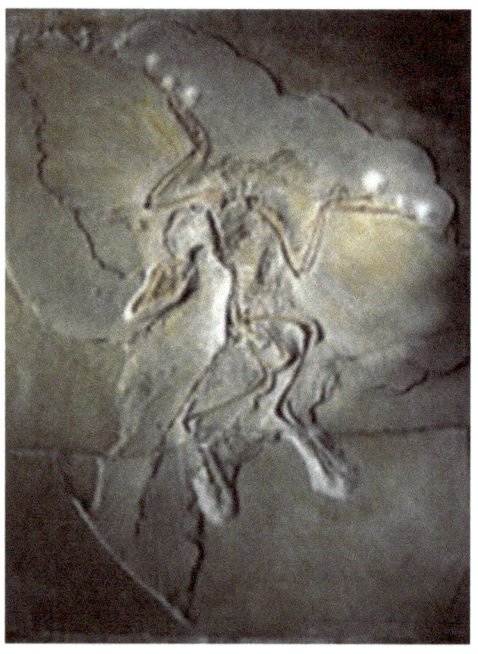

Archaeopteryx is one of the most famous transitional fossils and gives evidence for the evolution of birds from theropod dinosaurs.

We know today how and why mutations occur: the genome, the group of genes which, in every cell of a living organism, contains the genetic information to be transmitted to the offspring, is a dynamical system undergoing continuous changes because of the mobility of the genes and their property to detach from one chromosome and attach to a different position in the genome. The genetic information contained in the modified genome will be slightly different from the one in the initial genome. It can then happen sometimes that in the reproduction, the genetic information is transmitted not as an exact copy of the one from the parents but with some variation, which can produce a mutation in the structure of the newborn with respect to the parents' structure.

Mutations can be induced (due to the action of external physical or chemical agents, called mutagens) or spontaneous (in the absence of known mutagens).

Spontaneous mutations are caused either by internal chemical factors or by errors in the processes on the genetic material; for instance, a base modified by the removal of an hydrogen atom, a modified nitrogenous base, the formation of a nucleotide without a base, errors in the replication, recombination or DNA reparation processes.

For the induced mutations, the physical mutagens are mainly radiations (X-rays, gamma-rays and ultraviolet), the chemical ones belong to several classes of compounds. The damages produced by mutagens include the substitution of bases with molecules having a structure similar to the ones usually present in the DNA but forming different and therefore 'wrong' couplings, the generation of molecules with the 'wrong' coupling properties, the disruption of the nitrogenous bases, the insertion or deletion of some base.

An important difference between physical and chemical mutagens is that the first act independently from the organism, while the effects of the chemical mutagens can depend on the biological system. While a radiation directly hits the genetic material, the interaction of a chemical compound with other molecules in the cell can change its characteristics.

A more detailed discussion of the genetic code and mutations will be given in a later chapter.

We hear or read sometimes that Darwin's theory is the 'theory of chance', this is completely wrong. The only 'casual' element in the theory of evolution is the mutation: this is obviously unforeseeable but once it happens, the course of evolution is not casual anymore, it follows the iron laws of natural selection.

I would like to conclude with one last consideration on the organs of great complexity and perfection, an issue—the eye in particular—so dear to creationists.

There is an organ even more perfect than the eye, whose surprising structure was not known to Darwin and natural theologists: the bat's ear.

To move and locate its prey in the dark, the bat has developed (natural selection built for it) a sonar system, based on the acoustic waves collected by the ear, millions of years before man invented 'sonar' and 'radar', so perfect to arise wander and admiration in any engineer building sonars and radars. The bat 'sees' the world through the echoes, the return to its ears, after reflection from surrounding objects, of the ultrasound emitted while flying.

The perfection of the system is in the solution found for several problems. Frequency and intensity of the emitted sound increase when the prey is 'in sight', allowing for a more accurate localisation and better detection of the prey's movements. The ear is opened and closed with the same frequency of the emitted sound (up to 200 times per second!), to avoid being damaged by the same emitted sound. A similar solution has been 're-invented' by radar builders during World War II to avoid damaging the antennas when the radar emitted high intensity signals. During flight, the ear receives echoes not only from the prey but from surrounding objects as well, including the sounds emitted by other bats. And yet it has been shown that bats are able to 'select' the echo due to the prey.

A detailed discussion of this topic can be found in Dawkins' book *The Blind Watchmaker.*

In the same book, there is a discussion of all the weak points in the various attempts to introduce corrections to Darwin's idea, like the 'punctuation theory' of Eldredge and Gould or Dover's theory of 'molecular drive'.

Here in conclusion, we recall the two 'pillars' at the basis of the Darwinian mechanism of natural selection: gradualism and time. **Gradualism**: new individuals are born due to the gradual accumulation of *slight* variations, sufficient to provide the individual with *slight* advantages on other individuals of the same species, but small enough to avoid estrangement of the individual from other individuals of the same species, that is to be considered a 'monster' and isolated, or even eliminated, and having difficulties to breed anyhow, which would obviously check the evolution process.

Time: an obvious consequence of gradualism is the need of a time lapse long enough for the accumulation of small variations to produce 'visible' changes in the phenotype, that is a new species.

Darwin's natural selection, with the ingenious idea of mutations as the initial engine, is still today the only theory providing a satisfactory explanation of the complexity of the natural world and it reminds me of the words written by Schiller (for Columbus):

With genius nature rests in eternal union,
What one promises the other certainly delivers.

3. Life on Earth

The origin of life is one of the great problems stimulating the intellectual curiosity of man. [Linus Pauling]

The only life we know is the one on planet earth, therefore, leaving aside possible 'abstract' definitions of the term 'life', let us consider life as we know it on earth.

A consequence of Darwin's theory of evolution, confirmed, as we now know, by the fact that all living forms, animals, plants and bacteria share the same genetic code formed by 64 'words' of the DNA, each of three letters, is that all living beings originate from an unique progenitor, the first who developed this particular genetic code and is at the origin of the tree of life. In fact, the probability that an identical genetic code developed independently in diverse forms of life (even in just two!) is so small that it can be considered null.

The problem still open is how the progenitor is 'born', that is the origin of life.

The trivial answer comes from the road traced by St Augustine of Hippo:

...there is no need for a deep examination of the nature of things, as was done by those called physicists by the Greek... To a Christian it is sufficient to believe that the only cause of

all things created, celestial or terrestrial, visible or invisible, is the Creator's goodness…and that nothing exists, except for He himself, whose existence does not derive from Him.

A Mosaic in the Cathedral of Monreale (Sicily) depicting God's Creation of Animal Species

The answer of Buddhism makes more sense:

In the search for truth, there are certain questions that are not important. Of what material is the universe constructed? Is the universe eternal? Are there limits or not to the universe? If a man were to postpone his search and practice of Enlightenment until such questions were solved, he would die before he found the path. [Buddha].

Or, in other words, it is useless to reflect for a lifetime about the origin of life if the true objective of man is the cessation of suffering (Enlightenment).

Till the seventeenth century, it was believed that life can arise 'spontaneously' from inert, lifeless matter; this was the theory of 'spontaneous generation' or abiogenesis. God created man and big animals directly, all 'inferior' animals (like worms and insects) and other microscopic forms of life (like fungi and bacteria) are born spontaneously from rotten carcasses or decomposing organic substances.

This theory goes back to Aristotle and his authority contributed to keep it intact for centuries, the theory is based on common observations of 'spontaneous birth': earthworms emerge from the ground, rats appear suddenly in deposits of corn, white silkworms and flies appear on rotten meat, micro-organism swim in a rotten meat broth. All observations, except of course for the last one, certainly known to Aristotle.

The experiments by Francesco Redi in 1668, Lazzaro Spallanzani in 1765, and the conclusive one by Louis Pasteur in 1864 proved such theory groundless. This had its roots in the observation of phenomena while Redi, Spallanzani and Pasteur, followers of the new scientific method, performed experiments to study those phenomena 'eliminating the external causes', in this case, the access of flies or other animals to the rotten carcasses or the presence of micro-organisms in the initial organic broth.

Redi observed that maggots and flies appear on a piece of meat placed in an open jar, but not if the jar is sealed. In other experiments, he sealed the jars of the test group with gauze, which prevented flies to enter but allowed the circulation of air, to make sure that the absence of air in the sealed jars was not the reason why the maggots did not develop. Again, no maggots appeared, thus confirming the result of previous experiments.

From these results, he deduced that flies could be generated only by other flies in the open jars, where they could deposit their eggs on the meat, which was impossible to do in the sealed jars.

To study the spontaneous generation of micro-organisms, Spallanzani first and then Pasteur performed experiments with organic 'soups' which, when left to decay in the air, revealed the presence of micro-organisms when examined with a microscope. But when the soups were left to boil for times long enough to kill any micro-organism already present in the soup and then the containers were sealed, the soup was sterile, micro-organisms did not grow. They appeared only if the soup did not boil for a sufficiently long time or if the sterile soup was again in contact with air, which proved that micro-organisms can also be transported by air.

These experiments proved that, as flies can only be born from flies, micro-organisms can only be born from micro-organisms. In the words of Pasteur: *The theory of spontaneous generation will never again raise from the mortal blow inflicted by this simple experiment.*

The problem of the origin of life is then open not only for man but for all living beings. But 'origin of what'? To grasp the essence of the problem it is necessary to understand the characteristics of life (on earth).

WHAT IS LIFE (on earth)

Living beings are 'chemical machines' with the following fundamental properties:

- **All** living beings are made for 99% by Carbon (C), Hydrogen (H), Oxygen (O) and Nitrogen (N).
- **All** living organisms are constituted by cells, from single-celled bacteria, to the billions of cells of man. Cells have the property to reproduce themselves using the 'genetic code', identical for all the cells of all living beings.
- **All** living beings are constituted by proteins and nucleic acids which, in turn, in **all** living beings are constituted by a finite number of smaller molecules: twenty amino acids for the proteins, four nucleotides for the nucleic acids.
- **All** the organic compounds entering in the composition of a cell contain carbon: the activities of a cell depend therefore from the chemistry of carbon and its compounds, called organic chemistry. The carbon atom, given its particular structure with four valency electrons, has an unique 'easiness' to form bonds with other atoms and can therefore form more compounds than any other element. For instance, Lithium can form one compound with hydrogen atoms, Oxygen-two Nitrogen-seven...Carbon more

than 2300. This 'sociability' of carbon is fundamental to understand the organisation and chemistry of life.

What clearly emerges from these basic characteristics is that the enormous variety of life forms on earth present a unitary design at the microscopic level, which suggests a descendance of all living beings from the first organism who developed such characteristics.

To understand how to formulate the problem of the origin of this first organism, it is necessary to introduce some indispensable elements about the 'chemistry of life', the interested reader can find more extensive discussions in specialized textbooks, like the ones by Asimov, Adler and Monod (see the bibliography).

Proteins are 'giant' molecules formed by amino acids, which are molecules whose basic elements are Carbon (C), Oxygen (O), Hydrogen (H) and Nitrogen (N), and contain both the 'amino group' (NH_2) and the 'carboxyl group' (COOH). The known number of amino acids is 20, common to all living beings, from bacteria to man, pointing once again to the unity of all organisms at the molecular level.

Each protein is made of several different amino acids, up to tens of thousands, with the amino acids bound in chains, called 'peptides', which form because the carboxyl group of each amino acid is 'chained' to the amino group of its neighbour with a bond called 'peptide bond'. Adler (*How Life Began)* compares an amino acid to a person, his left hand representing the amino group and his right hand the carboxyl group, and a peptide to a chain of persons with the left hand of one holding the right hand of another ('peptide bond'), so that each has a free hand to hold the hand of another person to form long chains.

We know today that proteins are 'polypeptides', formed by long chains of peptides (hemoglobin for instance, an average size protein found in the blood, contains 550 peptides), and all 20 amino acids are present, except for some exceptions in particularly simple proteins. A fundamental property of proteins is their 'stereo-specificity', that is their ability to 'recognise' other molecules from their form.

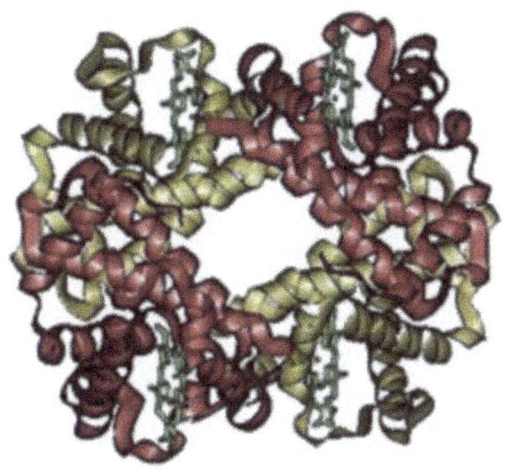

Structure of Haemoglobin

It is this particular property that allows a protein to 'discriminate' and 'choose', among the molecules in its surroundings, those needed for the process overseen by that protein.

Living beings are chemical machines which 'build' themselves absorbing from their surroundings elements like oxygen, water, carbohydrates, minerals, fats, sugars, solar radiation, etc. and transforming them in tissues, organs, energy, etc. is the set of life-sustaining chemical reactions in organisms. The conversion of the energy in food to energy available to run cellular processes; the conversion of food to building blocks of proteins, lipids, nucleic acids, and some carbohydrates; and the elimination of metabolic wastes. These enzyme-catalyzed reactions allow organisms to grow and reproduce, maintain their structures, and respond to their environments.

While the proteins are formed by chains of peptides, nucleic acids are macro-molecules formed by chains of nucleotides. A nucleotide is a molecule formed by a nitrogenous base with the same elements of amino acids (C, O, H, N) but a different structure, bound to a sugar and to a group containing phosphorus (P). In the nucleotides of all living beings, there are only four nitrogenous bases, named adenine (A), guanine (G), cytosine (C)—not to be confused with the symbol for carbon—and thymine (T). We shall discuss later their basic importance; they are the four 'letters' constituting the alphabet of the genetic code. The nucleic acids differ among themselves for the kind of sugar, which can be either ribose or deoxyribose. The nucleic acid containing ribose is called ribonucleic acid (RNA), the one containing deoxyribose is called deoxyribonucleic acid (DNA). To be more precise, in RNA, the thymine is replaced by uracil (U), a molecule with similar but not identical structure.

As proteins are polypeptides, DNA and RNA are polynucleotides. The DNA is formed by two 'complementary' chains of polynucleotides curled to form a double helix, with the two chains constituted by the sugar phosphate of the nucleotides (like the bars of a ladder or backbone) wound as an helix around a central axis and the nitrogenous bases protruding towards the interior binding to each other (the ladder's rungs): adenine has a spontaneous tendency to form a bond with thymine (A-T) and guanine with cytosine (G-C).

In living beings, proteins and nucleic acids are found inside cells. These were discovered only after the development of the microscope, when it was observed that living tissues are constituted by 'small cavities' filled with liquids, like those observed in 1665 in the structure of cork by Hooke, who named them 'cells' in analogy with the cells of a monastery. The name was then extended to the small cavities in living tissues and by mid-1800, it was clear not only that all living matter is formed by cells, but also that every cell is an independent living organism, as proved by the existence of micro-organisms formed by a single cell.

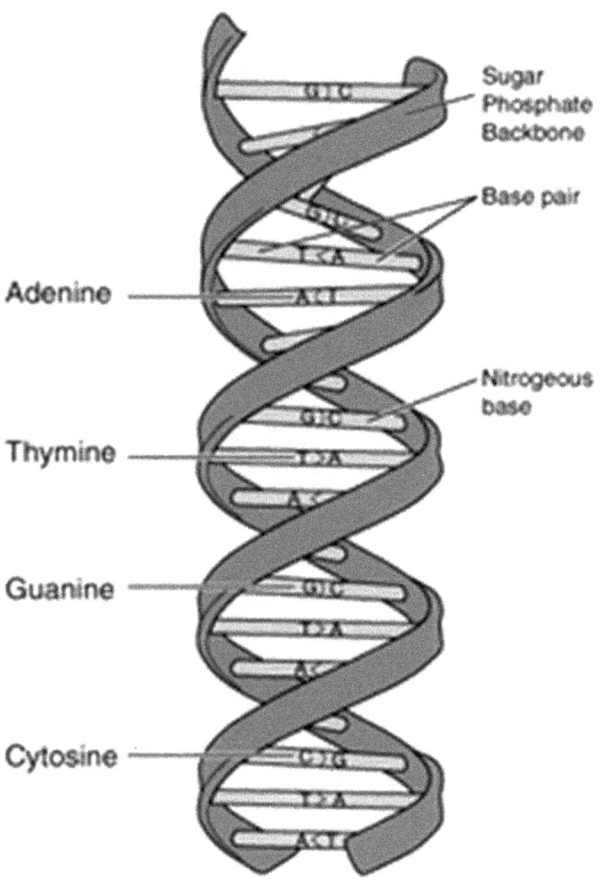

The key structural features of the DNA double helix.

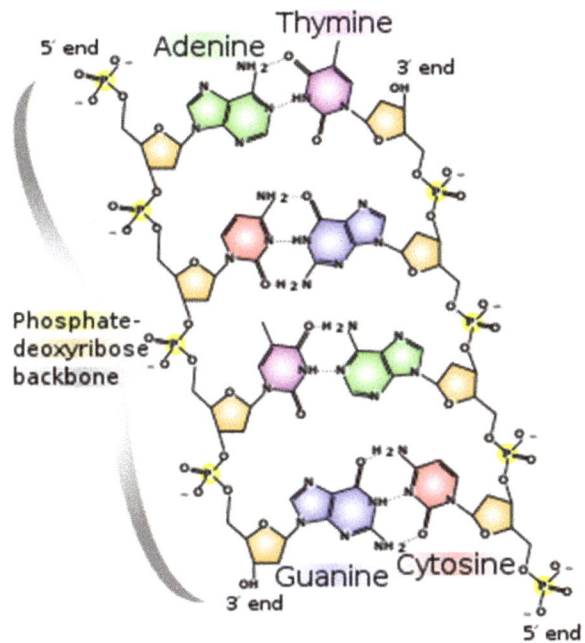

Chemical Structure of DNA

All animal and vegetal cells contain the same liquid, called cytoplasm, formed by proteins and RNA. Except for bacteria and blue algae, called prokaryotes, the cells of all living beings (eukaryotes) contain a 'nucleus' formed by proteins, DNA and RNA. The DNA is contained in filaments inside the nucleus called chromosomes, each filament can contain several millions of nucleotides, for instance, the largest human chromosome contains 250 million pairs of bases. The number of chromosomes is fixed in the eukaryotic cells of every animal and vegetal species. The DNA in the prokaryotic cells may or may not be organised in chromosomes, in bacteria there is a single DNA molecule.

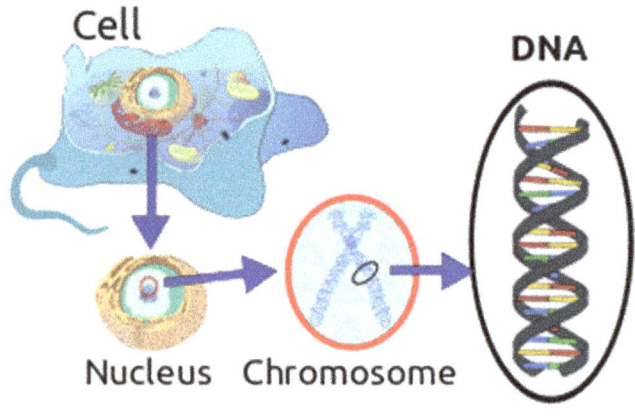

Location of eukaryote nuclear DNA within the chromosomes

The DNA is the main agent in cellular division (mitosis) and inheritance of characters. In the sexual reproduction in mammals, from the union of an 'egg cell' with a spermatozoon contained in the seminal liquid, a fertilised egg is formed, from which, after repeated divisions, a new individual is formed.

In the division process of a cell, the number of chromosomes is doubled, each chromosome replicates itself, the original chromosomes and their copies are then pulled to opposite sides of the cell, forming two genetically identical nuclei with the same number of chromosomes of the 'mother' cell. The new cell then divides to form two genetically identical 'daughter' cells, the process is then repeated in each 'daughter' cell and so on up to the formation of the multicellular organism. This division process however does

74

not take place in the formation, with a different process called 'meiosis', of the egg and sperm cells, which then contain only half of the chromosomes present in all the other cells of the organism: when they unite, the fertilised egg has then the full number of chromosomes, half from the father's spermatozoon and half from the mother's egg cell and this full number is then transmitted, through mitosis, to all the cells of the new organism (except of course to spermatozoa and egg cells). In asexual reproduction, common to the vegetable world and to some animals with a primitive structure, like sponges and starfish, the chromosomes of the new individual come from only one of the parents.

The DNA replication process is due to the double helix structure and the 'binding rules' of the nitrogenous bases: A binds T and G binds C. When the two filaments separate as the process begins, each one builds a new filament using the surrounding material, 'choosing' the bases of the replicated filament according to the binding rules: the result is a 'complementary' filament which can then join its 'parental' filament to reconstruct the double helix so that the 'daughter' cells are born with chromosomes identical to the ones of the mother cell.

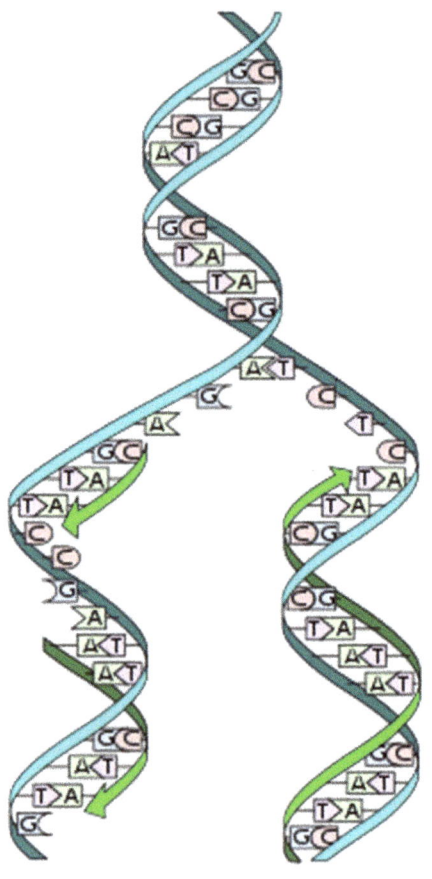

DNA replication: The double helix is 'unzipped' and unwound, then each separated strand (turquoise) acts as a template for replicating a new partner strand (green). Nucleotides (bases) are matched to synthesise the new partner strands into two new double helices.

The next task, after replication, is the construction of proteins, of all the types of proteins needed for the formation of the new individual: organs, tissues, brain…all must be built following the 'information' contained in the parents' chromosomes, that is inherited from the parents. Where is this 'information', how is it transmitted and used? Mendel's experiments, once repeated and extended, had shown that the hereditary characters, which determine the observable characteristics of an organism (the phenotype), are transmitted whole, part from the father and part from the mother, for generations. These characters, which were named 'genes', were therefore immediately associated with the chromosomes. But given the limited number of chromosomes and the enormous number of inherited characters (in humans, for instance, there are 23 pairs of chromosomes and thousands of inherited characters), the genes cannot be the chromosomes themselves, each chromosome must be a collection of genes.

We know today that the gene is the DNA molecule, and a chromosome can contain millions of genes, we also know that there is a 1:1 relation between genes and proteins, in the sense that each gene builds its own particular protein. To do it, the gene must 'dictate' the sequence of the hundreds or thousands of amino-acids which form that particular protein and not a different one, for any position in the chain of polypeptides it must 'choose', between 20 different possibilities, the correct amino acid. If the DNA molecule had 20 different 'characters', each one could be associated to one of the 20 amino acids, but the DNA molecule has only four different 'characters', the four nucleotides A, G, T, C, and therefore the mechanism cannot be a 1:1 correspondence between nucleotides and amino acids.

How can four characters determine the choice between 20?

The solution was found in 1954, not by biologists but by the English astronomer George Gamow, who suggested that the four nucleotides could form combinations corresponding to 'words' in a 'code' which associates each word to an amino acid.

If these 'words' were formed by combinations of two of the four nucleotides (for instance, AA, AG, TC, CA...), it would be possible to form only $4^2=16$ words, not enough to select 20 amino acids, therefore the minimum number of nucleotides needed to form a 'word' must be three (for instance, AAA, ATC, GCA, CCG...): in this case, it is possible to form $4^3=64$ 'words', largely sufficient. A triplet of nucleotides (a word of the code) is named a 'codon' and contains the command for the construction of a particular amino acid. For instance, the thymine repeated in a triplet of uracils (UUU) encodes the amino acid called phenylalanine. It should be noticed that since the codons are 64 and the amino acids 20, some amino acids can be encoded by more than one triplet, but the opposite is not true: to each triplet, corresponds only one amino acid.

The DNA and RNA molecules can then be considered as long chains of codons which, all together, constitute the 'genetic code' of any individual. The genetic code is the ensemble of the rules through which the information coded in the nucleic acids forming the genes is translated to build the proteins in the cell. This genetic information, that is which amino acids to choose and how to assemble them to reach the desired polypeptide chain (protein), is given by the order in the sequential disposition of the nucleotides. The genetic code is read 'without punctuation', that is linearly triplet after triplet with no superposition (for instance, the last base of a codon cannot be read as the first of the successive codon).

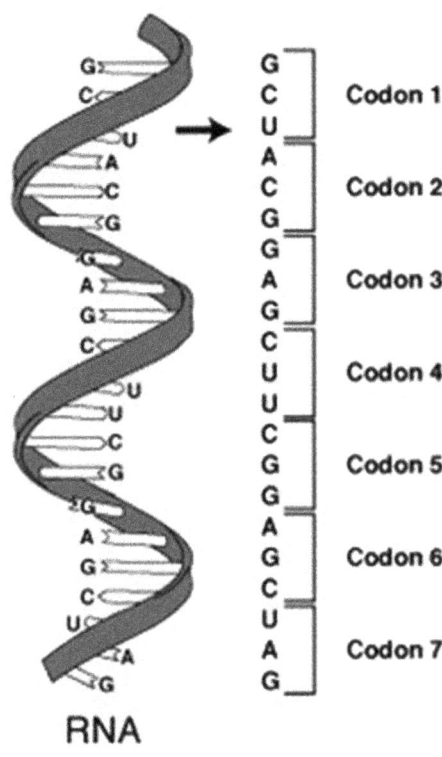

RNA

Ribonucleic acid

A codon is defined by the initial nucleotide from which the translation starts. For instance, the string GGGAAACCC, if read from the first position, contains the codons GGG, AAA, and CCC; if read from the second position GGA, AAC; if read from the third position GAA and ACC. Every sequence can therefore be read in different ways, each of which will give rise to a different sequence of amino acids. When a given sequence of codons has been read—and the corresponding protein built—a 'stop' codon gives the command to quit reading. The stop codons are UAA, UGA and UAG, they indicate the point, inside the gene, where the sequence encoding the corresponding protein ends.

The 'construction' of proteins takes place in the cell cytoplasm, which is rich with the basic ingredients to be used following the orders from the genetic code. There are organisms (autotrophic) which build by themselves the amino acids in the cytoplasm, others (heterotrophic) which absorb them from food. The 'transcription' and the 'translation' of the information from genes to cytoplasm involve the RNA molecules through a series of processes too complicated to be discussed here. For our present purpose, it is sufficient to know that all these processes are just chemical reactions which take place spontaneously.

The inheritance of characters is ensured by the DNA duplication happening so that the daughter cells have the same genetic information of the mother cell. The DNA molecule contains the genetic information allowing all living organisms to function, grow and reproduce. We still do not know the moment when, in the history of life, the DNA assumed such fundamental role. May be the first nucleic acid used by living beings was rather the RNA, which also has the property to replicate itself. Since it has a simpler structure than DNA, only one filament, it is likely to have been at the basis of the first cellular metabolism.

For our purpose, it is also important to know that, in the DNA replication process, errors in the 'copy' of the second filament can occur, resulting in a 'daughter' genome modified with respect to the 'mother' genome, which produces the appearance of mutations with possible impact on the observable characters of an organism (phenotype). Darwin did not know it but used the idea anyway as the starting point of his theory, believing that one day, the origin of mutations would be known.

In conclusion, it should now be clear how to formulate the problem of the origin of life: how did the first 'replicant' form, the first structure able to make use of surrounding material to 'build' a copy of itself.

The Origin of Life on Earth

Darwin suggested that the original spark of life may have begun in a *warm little pond, with all sorts of ammonia and phosphoric salts, light, heat, electricity, &c., present, that a protein compound was chemically formed ready to undergo still more complex changes.*

The same idea is at the basis of one of the present hypotheses about the origin of life, the 'primordial soup'. In 1924, Aleksandr Oparin proposed that, in an atmosphere with little oxygen and with the energy from the solar ultraviolet radiation, organic molecules would be produced, which, accumulating in the sea, would have formed a primordial soup in which they could react forming ever more complex molecules, eventually forming droplets protected by membranes, like cells. These could in turn coalesce with other droplets to form more complex structures or divide into 'daughter' droplets, a sort of primordial metabolism. Oparin's important intuition was that the presence of oxygen in the atmosphere would have prevented this process, due not only to the ozone, which forms from oxygen, and acts as a screen for the ultraviolet radiation, but also to the oxygen high reactivity, which would have led to the destruction of the various structures as they formed.

Oparin's hypothesis has been developed and tested in the laboratory and it is today one of the most reliable theories to explain the origin of life. The organic compounds present in the oceans (or in little ponds) on the new-born earth, thanks to the 'sociability' of carbon, formed processes and structures of increasing complexity until, *in billions of years,* there appeared structures interacting between themselves and with the property to make use of the surrounding material (to 'feed') to produce copies of themselves. To interact and grow, these structures had to move, that is they had to be in a solution (rich with 'food') so that the molecules of the solvent, in their chaotic motion (Brownian motion), could push around—via collisions—the larger structures: once in a chaotic motion themselves in the soup, the latter had more chances to 'meet' other structures and react forming increasingly complex structures. On the earth, the solvent was water: the oceans or little ponds have been the 'primordial soup' in which life originated.

If so, the origin of life seems then to require the following conditions:

- The presence of liquid water.
- The presence of the elements C, H, O, N and of other elements, like phosphorus and iron, which are found in the proteins.

- Oxygen has to be present in its compounds but should not be 'free', that is, there should be no oxygen in the atmosphere.
- An energy source: chemical reactions leading to the formation of complex structures require energy.
- Time: the accidental mixing and reacting of the chemical products in the primordial soup leaves to chance (the chaotic motion) the possibility that the molecules needed to form a given structure come close enough to interact. Only on a very long timescale, there is a finite probability that, out of all possible 'combinations', sooner or later also the 'useful' ones have occurred.

Consider time first. The earth is born 4,5 billion years ago, the presence on the ground of liquid water, from the impact of meteorites and the precipitation of the enormous amount of water vapour released in the atmosphere by the continuous volcanic activity in the primordial earth, has been estimated at 4,4 billion years ago [Wilde S.A. *et al. Nature 2001; 409: 175*].

How long afterwards was life born? After about two billion years, was the answer till the middle of last century, since no older fossils had been found. But in 1950, palaeontologists had the idea to look for fossils in chert beds, a mineral with the appropriate characteristics for the preservation of bacterial cells, and since then there have been many discoveries of older fossils (bacteria). So far, the oldest rocks containing fossil bacteria are dated at 3.5 billion years ago. Life in its simplest form, the prokaryotic cell of a bacterium, arose therefore 'almost immediately' after the appearance of liquid water. But the first certain evidence of multicellular life is dated 2.7 billion years ago, from the first traces of photosynthesis activity. For more than one billion years, bacteria have been the only form of life on earth.

It is important to keep in mind these periods of time measurable in billions of years, unmeasurably larger for our mental habit based on centuries and millennia. On such enormous timescale, even the most improbable event can take place, if it does not contradict the laws of physics.

But the bacterium, though the simplest living being, is still an extremely complex structure. In the search for the first 'replicant', we must start from more elementary structures, the first fundamental step must have been the production of relatively simple organic molecules, like amino acids and nucleotides, the building blocks of life, starting from inorganic compounds present in the chemical-physical scenario of the young earth.

The solar system and the sun have been formed from the same cloud, which means that all the elements that we observe today on the sun were present on the earth, from hydrogen to iron, mostly in the form of chemical compounds. All the basic elements listed above, H, O, C, N, were therefore present in the form of compounds like ammonia, methane, and water vapour. An energy source was also present, the solar radiation, in particular the ultraviolet radiation, not screened by the presence of ozone in the atmosphere.

The question is: is it possible that organic compounds form out of these inorganic ones in the presence of a source of energy?

The answer is yes, and it was demonstrated for the first time in 1953 with an experiment performed by Stanley Miller and Harold Urey. A vessel containing a gaseous solution of methane, ammonia, hydrogen and water vapour was exposed to electric discharges as energy source: in just one week, organic compounds, including some amino acids, appeared.

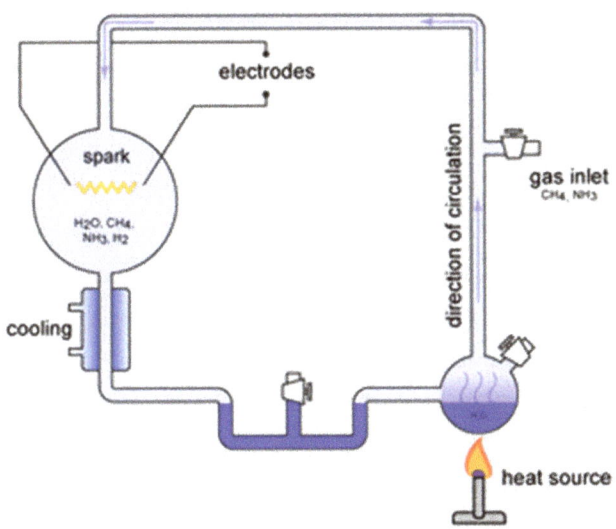

The Miller-Urey Experiment

Many similar experiments were done afterwards, adding to the Miller-Urey solution more inorganic compounds likely to be present on the young earth, like hydrocyanic acid and acetylene. They all proved the 'spontaneous' formation of sugars like ribose and deoxyribose, of the nitrogenous bases A, C, G, T and of chains of polymers.

Complex organic molecules have also been detected inside nebulae and meteorites, including traces of polycyclic aromatic hydrocarbons (IPA), the most complex molecule ever found in the universe. We can conclude that these transformations from inorganic to organic compounds are more a rule than an exception.

The first step towards life is then demonstrated: the building blocks of all living beings form 'spontaneously' in the chemical-physical conditions typical of the young earth. The accumulation of these first molecules provided a rich environment (primordial soup) for a successive chemical evolution.

The next steps are still not understood how from these initial building blocks the first 'replicant' and finally, the cell arose, capable to feed themselves, that is to use surrounding material to maintain their structure and reproduce.

But some little steps in this direction have already been made, in particular, it has been shown that: amino acids can spontaneously form small peptide, amino acids and small peptides can form closed spherical membranes, micro-spheres which could act as the cell's initial 'seed'. Some organic compounds, called phospholipids, can spontaneously form a double layer, a basic ingredient of the cellular membrane.

These steps all show that, once the building blocks are present, chemical evolution proceeds in the right direction.

Diverse theories have been proposed to explain the formation of structures of increasing complexity and the appearance of the first replicant. In the theory of the 'primordial soup' discussed above, these arise as a result of continuous collisions and interactions between simpler structures in perpetual motion in the oceans or little ponds, given that in a sufficiently long time, even the highly improbable combination of molecules forming a replicant becomes possible. When tossing a coin, even the improbable combination of 50 heads in a row sooner or later will be realised if we keep tossing the coin for a sufficiently long time.

Other theories have also been proposed; for instance, the one by the chemist Cairns-Smith based on the gradual formation of complex organic molecules on the surface of inorganic crystals in ponds of clay or mud, or the one based on the role of elements like iron and sulphur.

They all share time as the essential element: however, the first replicant formed, in the ocean, in ponds, in clay, deep under the earth's surface or in caves under the sea, it was a consequence of a chain of highly improbable events, made possible only by a very long stretch of time.

It should also be mentioned that a plausible hypothesis is that life formed not on the earth but in space and was then transported and sown on the earth by the impacts of comets and asteroids. Simple carbon compounds on the surface of comets, thanks to the energy from the solar radiation, have produced chemical reactions such that comets are covered with complex organic material, which may have 'rained' on the earth with the debris left by comets whenever they pass 'close' to us.

Complex organic molecules are present in nebulae and meteorites, may be originated from the same chemical reactions which we assume happened on the young earth, and since the universe has been around for billions of years before the earth was born, the time factor is all in favour of an extra-terrestrial birth of the first replicant, although this does not solve the problem of 'how' did it arise; it simply moves it elsewhere.

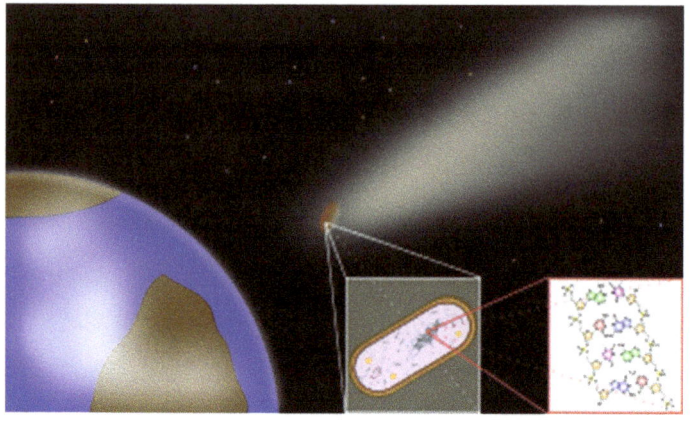

Panspermia proposes that bodies such as comets transported life forms such as bacteria—complete with their DNA—through space to the earth.

Conclusion. The origin of life is still a mystery, only the first step has been proved, the 'spontaneous' appearance, in the conditions of the primordial earth, of the organic molecules which form the basis of life (as we know it). The next steps, from the organic molecules to the first replicant, from the replicants to the prokaryotic cell, from the prokaryotic to the eukaryotic cell, from the single cell to multicellular organisms, are still at the stage of hypotheses. Given the enormous amount of time, millions or even billions of years, for each of these steps to be realised, obviously we will never be able to 'witness' the occurrence of any of them in the laboratory, as we did for the first step.

But the development of quantum computers, the new frontier in the technological applications of quantum theory, might offer some hope. If we will be able to 'simulate' in a quantum computer the necessary conditions for the occurrence of those steps, starting with a 'soup' of organic molecules with their chemical interactions, may be after a few human generations the computer will show the presence of replicants…

4. The Improbable Homo Sapiens

...The only surviving species is that of homo sapiens, an undeserved flattering label, that this only living species of hominoids attributed to itself with naïve presumption. [A Toynbee, *Mankind and Mother Earth*]

Following the history of life as told by fossils, let us consider the chain of events which, on the earth, led to the appearance of man. With the notable exception of the 'Burgess fauna' discussed later, fossils are the remains of the 'hard parts' (bones, teeth, shells...) of ancient organisms, since the 'soft parts' (skin, feathers, internal organs...) are normally destroyed in a short time.

It is convenient to make reference to the 'geochronologic' scale, a timescale built on the basis of the fossil records, which divides time in geologic 'eras', formed by 'periods', with the separation between eras characterised by 'extraordinary' events. The eras are:

- Palaeozoic: starts with the Cambrian period, about 570 million years ago and ends with the Permian period, about 225 million years ago.

- Mesozoic: from the Triassic period, about 225 million years ago to the Cretaceous period, about 65 million years ago.
- Cenozoic: from about 65 million years ago to the present day.

This scale was adopted when the oldest known fossils were those of multicellular animals with hard parts, which appeared between 530 and 520 million years ago in what has been named the 'Cambrian explosion' because these appear 'suddenly' on a geological timescale. Thus, the oldest era, the Palaeozoic, began with the 'start' of life in the Cambrian period, all the time before that, from the earth's formation to the Cambrian explosion, was named pre-Cambrian, but it has later been named Cryptozoic (hidden life) after the discovery of older forms of life.

There are three extraordinary events.

-The Cambrian explosion separates the Cryptozoic from the Palaeozoic.

-The Palaeozoic is separated from the Mesozoic by the first great mass extinction, about 225 million years ago in the Permian period, in which about 96 per cent of all marine species living in the Palaeozoic disappeared.

-Between the Mesozoic and the Cenozoic, about 65 million years ago, in the Cretaceous period, there was a second mass extinction, the event where the dinosaurs disappeared.

Mass extinctions, caused perhaps by impacting meteorites, are the first element of 'improbability' in the appearance of homo sapiens, if the particular ancestor from whom this species descends did not survive 'by chance' to these extinctions, we would not be here. But the history of life as told by the most recently discovered fossils, in particular the faunas of Ediacara and Burgess, seems to suggest the improbability of the appearance not only of homo sapiens but of all, or almost all, the life forms we know today, at least in the interpretation of the palaeontologist S Gould in his book *Wonderful Life,* dedicated to the fascinating story of the Burgess fauna.

The apparent lack of life before the Cambrian, followed by the sudden appearance of all the modern animal groups in the Cambrian explosion, was a problem that troubled Darwin himself, since the absence of 'ancestors' of these 'evolved' organisms is in open contradiction with an evolution proceeding with small steps for long times, as required by the theory of natural selection, and Darwin assumed that the reason was the imperfection of the fossil record: today they are not here, we will find them tomorrow…And we did find them.

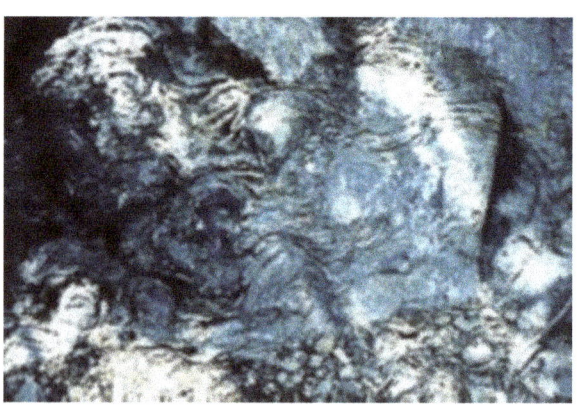

Pre-Cambrian stromatolites in the Siyeh Formation, Glacier National Park

The first form of life is found in the 'stromatolites', marine sedimentary rocks formed by the growth of layer upon layer of single-celled prokaryotic microbes, discovered in rocks aged about 3.5 billion years ago. If these are indeed the 'first' form of life after the earth's formation 4.5 billion years ago, the step from the first replicant to the simplest cell required about one billion years. These prokaryotic organisms have been the only life form on earth till the appearance, about two billion years ago, of the first single-celled eukaryotic organisms, like the amoeba and the paramecium. Notice that the step between the prokaryotic cell (without a nucleus) and the much more complex eukaryotic cell (with nucleus and cytoplasm) is so improbable that it took 1.5 billion years to complete. Then we must wait for about 200–500 million years before the first multi-cellular organisms appear: the first fossil algae are dated between 1.5 and 1.8 billion years ago.

Then, after hundreds of million years, the first 'animals' appear, the oldest fossils are dated between 620 and 550 million years ago. They have been found in several sites in the world but are collectively known as the fauna of Ediacara, the name of the Australian mountain where the best-preserved fossils of this kind have been found.

The Ediacaran fossils show a variety of different structures, for instance: radial, concentric symmetry, like jellyfish, with a radius of a few centimetres, like Cyclomedusa.

Fossil of Cyclomedusa-radial symmetry with hooked arms, like
Tribrachidium heraldicum

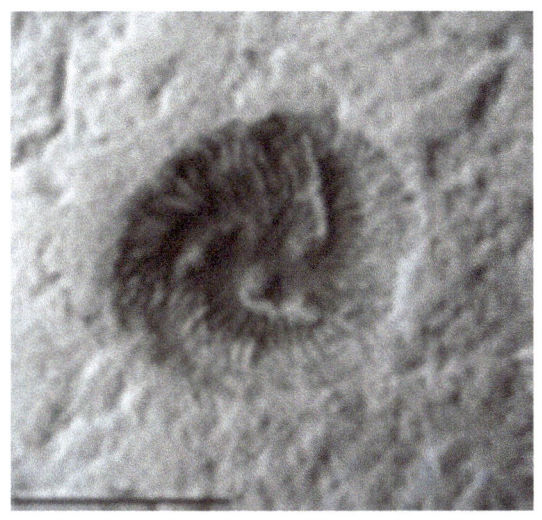

Fossil of Tribrachidium heraldicum

Egg-shaped or elongated forms, with a sort of horseshoe shaped 'head' and a body formed by a number of segments variable between five (Vendia) and more than forty (Spriggina).

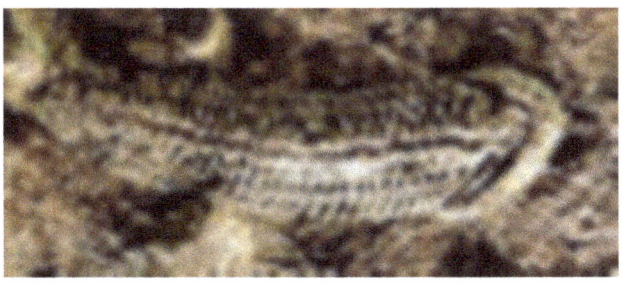

Fossil of Spriggina

Organisms with a bipolar symmetry, with a central furrow from which numerous segments spring out, like the Dickinsonia.

Fossil of Dickinsonia

Many organisms in the Ediacaran fauna have characteristics much different from those of 'modern' plants and animals and it has been suggested that they may represent an 'attempted evolution' of multicellular animals with no descendants: there are no traces of the circulatory, breathing and digestive systems in their bodies, may be the metabolic processes of these organisms took place directly through their bodily surface.

Apparently, the Ediacaran fauna disappeared shortly before the end of the Cryptozoic, with the start of the Cambrian, but the causes of such extinction are still not clear, a possibility is the occurrence of an anoxic event, that is a decrease or absence of oxygen in the levels just below the sea surface.

Before the sudden extinction, the Ediacaran fauna has been the dominant form of life for about 100 million years: the fossils 'disappear' about 543 million years ago and almost at the same time, between 530 and 520 million years ago, the first fossils of 'modern' animals make their appearance in the Cambrian explosion. Would the modern forms be born if an unknown event did not eliminate the Ediacaran fauna?

The same question, even more dramatic, arises from the study of the Burgess fauna, fossils from an age following shortly the Cambrian explosion, found for the first time in 1909 by the palaeontologist Charles Walcott in a rocky shale near mount Burgess, in Canada. Since then, similar deposits have been discovered in other places in the world, with similar fauna, therefore the Burgess fauna is not an isolated event, but it rather represents the presence all over the world of these kind of animals coexisting with the 'modern' organisms which appeared in the Cambrian explosion.

The great importance of these fossils is in that they contain soft parts, preserved from the moment when they were 'buried' in the shale. Walcott classified them as 'primitive forms' of existing phyla, but later studies have shown that many of these fossils belong to unknown taxonomic groups. Many animals in the Burgess fauna present anatomic characteristics never seen before, with only a vague similarity to known animals. The main examples of animals which did not fit any known bodily plan are Opabinia, with five eyes and a protuberance resembling a vacuum cleaner; Wiwaxia, with a look resembling a birthday cake with candles.

Opabinia

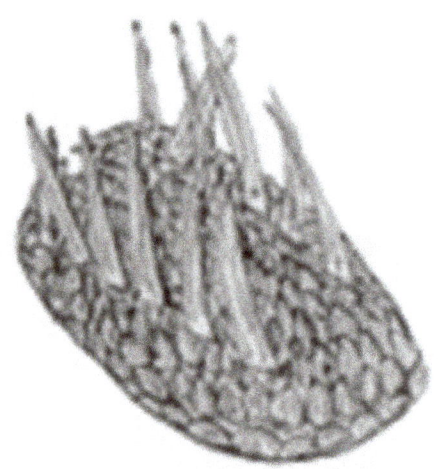

Wiwaxia

Hallucigenia, which seems to have no mouth and it is not yet clear if it was an animal or just a part of another animal.

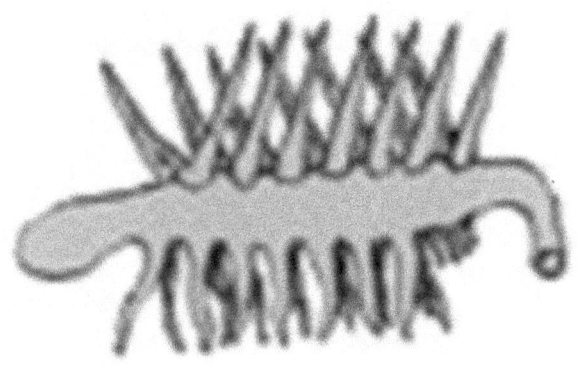

Hallucigenia

Anomalocaris, the largest predator in the Burgess fauna (with a length of about 60 cm, compared to the few centimetres of all others), found also in several different sites. The parts forming Anomalocaris were initially interpreted as separate animals, like jellyfish, sea cucumbers and shrimps, the story of their correct reconstruction as anatomical parts of a single animal is told in the cited book by Gould.

Reconstruction of Anomalocaris

The first complete Anomalocaris fossil found.

The Burgess fauna is also important because it contains an enormous variety of forms (anatomical plans), 'larger than those of all organisms living today in the seas of the Earth' according to Gould (*Wonderful Life*). As with the fauna of Ediacara, also with the Burgess fauna and with the other two mass extinctions in the Permian and Cretaceous periods, we are faced with an evolutionary history that 'eliminated' a large amount of 'possible' living beings.

In his classification of animals, the great French naturalist Georges Cuvier in 1800 introduced the concept of 'phylum', plural 'phyla', assigning all the animals with the same kind of bodily general organisation to the same phylum. The animals living today belong to a relatively small number of phyla, some examples:

Porifera (formed by a colony of cells with a porous skeleton, e.g., sponges), Coelenterates (with a 'bag' structure, e.g., jellyfish, sea anemones), Platyhelminths (flat worms, e.g., tapeworm) and Nematodes (with a round cross-section, e.g., hookworm), Mollusks (with a muscular foot and round mantle shell, e.g., clams, oysters), Annelids (segmented worms, e.g., earthworm), Arthropods (with segmented bodies, e.g., lobster, spiders, insects), Echinoderms (with radial symmetry, e.g., sea urchins, starfish), Chordates (with a backbone or dorsal nerve cord, e.g., man, snake, frog).

Arthropods form the most successful phylum in the evolution of life; this phylum contains many more species than all other phyla together.

The animals belonging to the phylum of chordates can have an internal cartilaginous axis (notochord), evolved into a backbone: the latter are the vertebrates, including fishes and mammals.

All the phyla of living animals were present in the fauna 'born' in the Cambrian explosion, but the Ediacaran and Burgess faunas contain many more phyla which did not survive. Why? Consider, for instance, the most successful phylum, the Arthropods: in the Burgess fauna, there are more than 20 distinct anatomic plans of arthropods, modern arthropods are all contained in just three basic anatomic plans, even though there are more than one million species of insects. The number of species increased during evolution but within a restricted number of anatomies compared to the greater possibilities that could have been realised if the other 17 anatomic plans, present in the Burgess fauna but not represented today, had not disappeared. It is as if—at the start—life could dispose of many channels; suitable for evolution but chose to exploit only a few of them. If these different anatomical plans had continued their evolution, may be today the largest predator in the seas would not be the white shark but some gigantic form evolved from the Anomalocaris.

One of the lucky survivors of the Burgess fauna was the Pikaia gracilens, a leaf-shaped animal, a few centimetres long. It was initially classified by Walcott as a worm, but later recognised as a chordate, belonging therefore to our same phylum. Pikaia is one of the oldest chordates, the ancestors of vertebrates, the fossils of true vertebrates belong to a much later age after the Burgess fauna.

Pikaia gracilens

Perhaps Pikaia is not our only ancestor who survived, but the absence of other fossils of chordates from the age of the Burgess fauna probably means that the chordates were not a very successful phylum and the survival of Pikaia can be symbolically considered as a 'winning ticket' for homo sapiens in the 'lottery' of life.

The animals with 'modern' phyla from the Cambrian explosion have replaced the Burgess fauna and for more than 100 million years, life has been confined to the seas; the first organisms to colonise the emerged land were plants, about 400 million years ago, followed after a few million years by arthropods, mollusks and worms. After some hundreds of million years, appear on land the first 'lunged' fishes, emerging from the sea to exploit the atmospheric oxygen, may be due to a lack of oxygen in areas of saline water, developing 'bags' to store the inhaled air, later evolved into lungs. From these, first colonisers of the emerged lands, evolution has then produced the amphibia 300 million years ago and the reptiles at the start of the Mesozoic period. The first mammals and birds appear only after the first great mass extinction at the end of the Paleozoic.

In the Permian–Triassic mass extinction, about 251.4 million years ago about 81% of marine species and 70% of species of land vertebrates disappeared as deduced from a statistical analysis of strata rich with fossils. This was also the only known extinction of insects.

Trilobites were highly successful marine animals until the
Permian–Triassic extinction event wiped them all out

More than 280 out of 329 genera of marine invertebrates
disappeared, the largest part of the later fossils of insects
significantly differs from those living before the extinction,
more than two-thirds of the families of terrestrial amphibia
and reptiles went extinct, the great herbivorous suffered the
largest losses.

After the extinction, the lystrosaurus, a herbivore reptile
with the average size of a pig, was by far the most widespread
terrestrial vertebrate, constituting 90% of the terrestrial
vertebrate fauna.

Lystrosaurus

The archosaurs, ancestors of dinosaurs and crocodiles, started to prevail, and evolved from a group of archosaurs becoming the dominant group in the terrestrial ecosystem until they disappeared in the Cretaceous mass extinction, clearing the way to the evolution of mammals. It is estimated that in this second great mass extinction, 75% of living species disappeared with the dinosaurs.

The mechanisms proposed to explain the mass extinctions range from gradual changes of the habitat (the level or amount of oxygen in the seas, increased drought, climate change) to a catastrophic event (collision with a meteorite, increased volcanic activity).

The evidence of collisions occurred at the same time of these events include meteoritic fragments in Antarctica and grains rich with iron, nickel and silicon, which could have originated from an impact, as well as several impact craters, along the north-western coast of Australia, in eastern Antarctica and the Gulf of Mexico itself. But the hypothesis that an impact was the main cause has not yet been proved.

Two large-scale volcanic events in the Permian period have been confirmed, one in China in an area near the equator, the other one in Siberia where one of the largest volcanic events ever happened on the earth, a surface of more than two million km^2 was covered with lava. The volcanic eruptions could also have had three consequences:

-the screening of solar light by the dust clouds produced may have reduced the photosynthesis, both on land and in the seas, generating a collapse of the food chain.

-acid rains may have killed terrestrial plants, mollusks and plankton.

-the emission of carbon dioxide may have caused global warming and a decrease of oxygen in the seas.

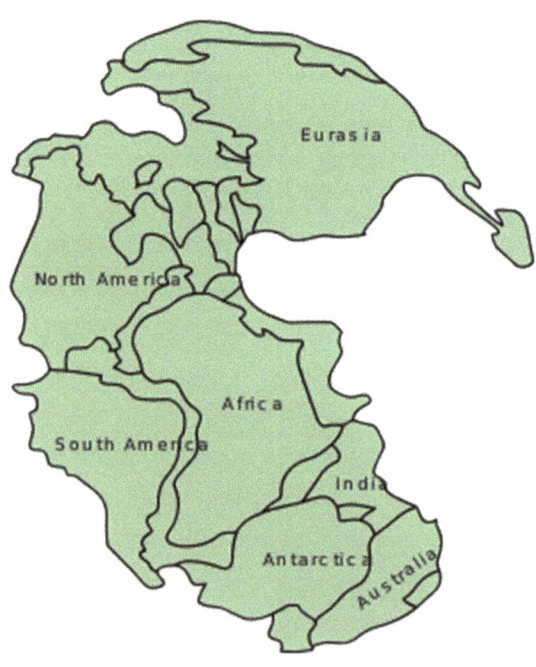

Map of Pangaea with modern continental outlines

Finally, the effects produced, around the middle of the Permian period, when all the continents were united to form the super-continent Pangaea, should be considered. This had three consequences:

-the substantial reduction of the superficial aquatic environments, the most productive areas of the seas.

-the free circulation of animals, causing an increased competition with invading organisms.

-the alteration of the oceanic and atmospheric circulation, producing heavy rainfall near the coasts and drought in the internal areas of the super-continent.

The formation of Pangaea has therefore probably determined extinctions of fauna, mainly marine, but it does not seem possible that it was the only cause of the 'great blight' of the Permian.

How to read the history of life in the light of the mass extinctions?

According to Darwin's theory, the 'losing' species in the fight for survival form 'dry branches' on the tree of life but should we interpret mass extinctions as the extinction of the losers? The answer is probably 'no' for two reasons: natural selection requires long times, probably much longer than the timescale of mass extinctions and there is no evidence that those who disappeared were not as fit for survival as the rest. Dinosaurs were certainly perfectly fit to be winners.

Mass extinctions were therefore mostly due to 'external' causes, deep and sudden changes of environmental conditions, due for instance to increased volcanic activity and impact of meteorites, changes that did not leave to living beings the time to adapt to the new conditions and continue their slow evolutionary path. A path that started again only for the few 'lucky' survivors.

If so, survival after the episodes of mass extinctions is entirely due to chance, while the theory of natural selection has no room for chance, except for the initial mutation. The interpretation of mass extinctions as the casual survival of few 'lucky' ones from the 'decimation', attributes to the role of chance also the emergence of the life forms as we know them today. In the words of Gould: *if run again, the movie of life would give a substantially different ensemble of surviving anatomies and a following story which would make sense perfectly though being completely different from the one we know* [S Gould, *Wonderful Life*].

According to this 'theory of contingency', evolution, up to homo sapiens, has been determined by the combination of the iron laws of natural selection and a series of casual events extremely improbable, in the sense that if the story of the evolution of life on earth started again from the beginning, for instance from the first prokaryotes, the probability that all the events leading to homo sapiens would be repeated is vanishingly small.

This should be kept in mind when discussing life on other worlds. As we shall see in a later chapter, the probability that life (the first replicant) formed or will form on other planets, be it Mars or one of the billions of planets external to the solar system, is very large, almost a certainty, but the probability that it evolved or will evolve in forms like those on earth is vanishingly small. Man is a 'unique' event in the universe.

An event emerging from a path of about ten million years, from the separation from the other primates to the appearance of the first forms of art, the pictures on rocks. A path that for millions of years has been almost exclusively morphological evolution, skull and brain particularly; only in the last 40.000 years a cultural development began, the first traces of man's creativity appeared. The ancestors of man appear for the first time in Africa where, between six and ten million years ago, they separated from the common ancestor with chimpanzees and gorillas. The first fossil records belonging to a primate who walked upright are dated about four million years ago, the famous Lucy, scientific name Australopithecus afarensis, a small creature about one meter tall and with thin bones. About three million years ago, this small creature became taller and more robust in the fossils of Australopithecus robustus and Australopithecus Africanus. The following fossils belong to a hominoid with a larger brain and hands with opposable thumb, which allowed him to make use, for the first time, of rudimentary stone tools, thereby named homo habilis.

This keeps evolving enlarging body and brain and developing, about 1.700.000 years ago, into a new species, homo erectus (this is not the first man to walk upright, Lucy did it millions of years before, but the fossils of this species were discovered long before Lucy).

Up to this point, our ancestors lived in Africa, where still live chimpanzees and gorillas from which they separated. "Homo" erectus leaves Africa about one million years ago and diffuses in the near and far east and in Europe, still developing body and brain till when, about 500.000 years ago, fossils show a structure more similar to modern man and such marked differences from homo erectus that it can be considered a new species, homo sapiens.

In millions of years, from the time it separated from chimpanzees and gorillas, there was a constant change in the anatomy, but Homo sapiens, even though reaching an almost modern form, is still far away from the 'cultural' characteristics distinguishing man, he still makes use of rough stone tools, while bows and arrows, houses and wall pictures are still hundreds of thousand years away.

Reconstruction of Australopithecus afarensis (Lucy)

Only 100.000 years ago, appear the first fossils witnessing the cultural progress of two descendants from homo sapiens, the man of Neanderthal in Europe and western Asia, men with a modern anatomy in Africa. But it is still a minimum progress, these are still 'stone age' men, no sign yet of art or adornments.

Then, about 40.000 years ago, with a sudden and deep change, in Europe and Africa makes its appearance the man of Cro-Magnon, named from the place in France where the first fossils were discovered. The Cro-Magnons are anatomically fully modern, they know the use of fire, cut stones to form sharp tools, build new tools combining stone and wood, like hatchets and lances, needles (which suggest the habit to sew their garments), hooks and nets, ropes, bows and arrows, the first 'houses'. And there appear the first signs of 'humanity': pictures on rocks, amber and obsidian jewellery. Thanks to the new weapons, for the first time appear in fossil sites, the bones of large animals, like the buffalo in the African sites and the bison in the European sites: Cro-Magnons had a nutritious diet hunting large and dangerous prey, not easy to kill for their unarmed ancestors. The dark side of the coin is that these successful hunters probably caused the extinction of many species, like the mammoth in North America, the giant deer in Europe and the giant kangaroo in Australia.

Almost at the same time as the Cro-Magnons appear, the Neanderthal man disappears. May be one evolved into the other, but it is more likely that this last form of homo sapiens evolved in Africa, invading later Europe and Asia and leading the Neanderthal to extinction, given their net superiority.

For other thousands of years, homo sapiens progresses towards modern man, traces of human activities appear, like burial of the dead and care of the ill, life expectancy increases from about 40 years of the Neanderthal to 60 years.

Reconstruction of a Neanderthal man

Then, about 10,000 years ago, appear the first traces of agriculture, animal domestication and stable settlements, evolution proceeds at a vertiginous pace compared to the slow one of the first million years, writing is invented, and history begins.

It took millions of years to reach the ancestor common to homo sapiens and the anthropomorphic apes, then in a few thousand years homo has built Notre Dame, painted the 'Gioconda' and reached the moon.

The other primates are still more or less as they were when our branch separated. Why? What is it that made the difference? The answer adds one more element of improbability to the appearance of homo sapiens, in addition to those discussed above in the evolution of life.

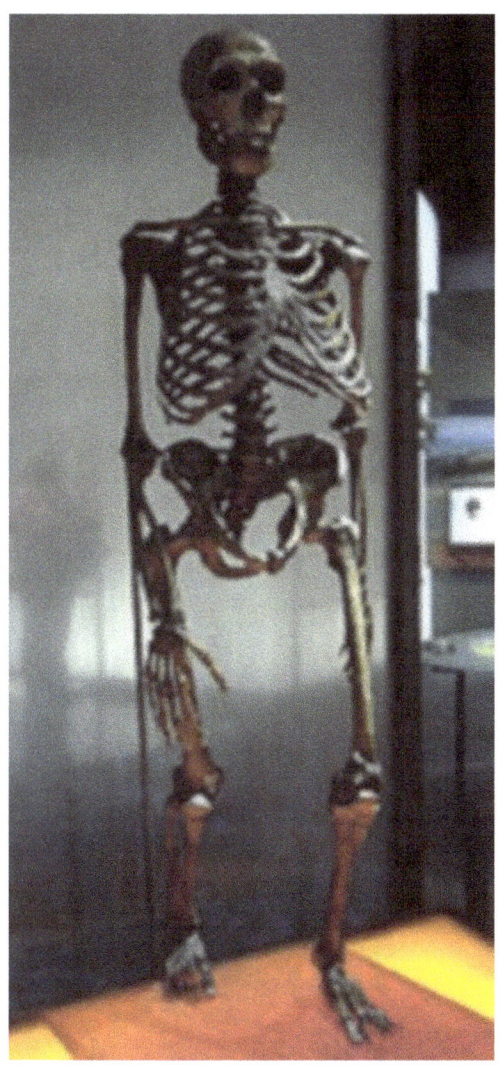

Skeleton of a Neanderthal man

Rock pictures from about 16.000 years ago

Forensic facial reconstruction of a Cro-Magnon

But first we must understand 'how much' we differ from chimpanzees and gorillas.

We have seen how each living organism is 'built' following the instructions of the genetic code, a dictionary of 64 words of DNA, three letters each, and the genetic code is universal; every living being, however different its form from the others, is built by the same genetic code, inherited from the ancestor common to all living species, the 'foot' of the tree of life. An organism is built when the information from the code is translated into proteins, each protein is a 'sentence', a chain of amino acid words from the dictionary. The differences between organisms are due to the fact that though using the same dictionary, different organisms build different sentences, in a given protein of two different organisms there are differences in the sequence of amino acids. These differences—the substitution in the chain of one amino acid with another—are casual, due to mutations, with a tendency to increase with time: the more two organisms depart from their common ancestor; the larger the difference between the words in a sentence, that is, there will be a greater number of different amino acids in the same protein of the two organisms.

Today biochemistry can compare the 'molecular sentences' of two organisms and measure their difference. For instance, using the protein 'cytochrome c' formed by about 150 amino acids and present in all organisms breathing oxygen, it has been found that human cytochrome c differs from the one of the kangaroos for ten amino acids, 21 from the tuna, 40 from yeast, one from the rhesus monkey.

Adding to the study of the differences between sequences of amino acids in proteins, the hypothesis that casual mutations, responsible for the evolution of genes, occur with a frequency almost constant in time, as proposed by Zuckerkandl and Pauling in 1962, it becomes possible to achieve a 'molecular clock' to estimate how long ago two species descending from the same common ancestor diverged. It was confirmed in 1963 by Emanuel Margoliash, with his studies on cytochrome c that the number of different amino acids of any two species is mainly due to the time since the two evolutionary lineages separated.

Such clock therefore allows to compare the times when the different branches of the tree of life separated, to achieve an absolute estimate of the time it is necessary to calibrate the molecular clock using the comparison with another clock, the radiometric dating used to estimate the age of fossils: the comparison of the two clocks provides an absolute timescale for the molecular clock. An example: if the radiometric dating of the tiger and lion fossils tells us that they separated from their common ancestor (becoming different species) five million years ago, and the molecular clock tells us that a given molecule, common to both, differs by 1%, we have a numerical correspondence between molecular differences and time. With this method, the molecular clock has been calibrated, the result is that the stable substitution of a *single* amino acid in the chain requires *seven million years*.

It has then become possible to evaluate the age of the different branches of the tree of life, that is, when two organisms separated. And we know today that bacteria and the first eukaryotes separated about 2.5 billion years ago, the progenitor of plants and animals lived 1.5 billion years ago, the common ancestor of insects and vertebrates lived one billion years ago. The tree of life is almost complete, the most beautiful illustration of Darwin's genius.

Before discussing what the molecular clock reveals about man, it is interesting to look at how the other clock works, the radiometric dating that allowed to establish the age of fossils and to order them according to a temporal scale.

Atoms with the same number of protons, which determine their atomic weight-and electrons, which determine their chemical properties, can differ in the number of neutrons in their nucleus, which contribute to their atomic weight, that is, they can exist in different 'isotopes' (meaning *same place* in the periodic table of the elements). A given isotope of an element is called nuclide and some nuclides (radioactive isotopes) are unstable, that is, they can decay to become a more stable isotope. The number of radioactive nuclides in any material therefore decreases with time, after a period called 'decay time' or 'half-life', characteristic of any radioactive isotope, half of the original radioactive isotopes have decayed. Knowledge of the law governing the decay and of the decay time allows then to use the abundance of particular nuclides in a material to calculate the time lapse from the moment when that material contained the original number of nuclides (provided we know it), that is the age of that material.

The comparison between the observed abundances of a radioactive isotope and its decay products forms the basis of radiometric dating, the most accurate clock to estimate the age of the earth and fossils. The estimate of such widely different ages, from the earth's billions of years to the hundreds of years of more recent objects, is possible using isotopes with decay times ranging from thousands to billions of years.

For instance, a radioactive isotope of Potassium, with a decay time of 1.3 billion years, and Uranium are used to find the age of billion years old rocks, with a (relatively small) error of a few million years.

The age of more recent objects can instead be found using an isotope of carbon, ^{14}C (carbon-14), which has a decay time of 5730 years, although this method is limited to about 50,000 years ago, because the concentration of ^{14}C in any material decreases fast and for older objects it is too low to be measured with a sufficient accuracy. On the other hand, its 'short' life-time allows to determine the age of 'recent' objects with an error of only few tens of years, that is why the ^{14}C method is the most accurate to estimate the age of 'historical' remains. A well-known example is the dating of the Holy Shroud, which was found by three independent laboratories to be the medieval work of a...joker, who certainly could not imagine the enormous success that his joke would enjoy for centuries.

The ^{14}C method allows dating materials of organic origin, like bones, wood, textile fibres, seeds, etc. Through respiration or photosynthesis, all living organisms continuously exchange carbon with the atmosphere (where carbon is present bound to oxygen in the form of carbon dioxide) and the use of this isotope is motivated by the fact that new ^{14}C is continuously produced in the atmosphere through capture of neutrons—present in cosmic rays—by nitrogen atoms in the atmosphere, so that an equilibrium is established between decay and production, keeping the ^{14}C concentration in the atmosphere constant.

If this did not happen, this isotope in the long run would disappear from the atmosphere and from living organisms.

But due to such equilibrium, an organism, while alive, absorbs oxygen from the atmosphere and continuously replaces the decayed ^{14}C so that the ratio between the concentration of ^{14}C and the concentration of other stable isotopes of carbon is constant, equal to the one found in the atmosphere. After death, the organism no longer exchanges carbon with the outside, the concentration of ^{14}C decreases with a known law and from the amount of ^{14}C present in the organic remains an estimate of the time of its death can be found. The American chemist WF Libby received the Nobel Prize in 1960 for the perfect tuning of this method.

The whole 'recent' history of homo sapiens has been unfolded by the combination of this dating method with the molecular clock. The result is quite surprising.

On the tree of life, man and chimpanzee separated between six and eight million years ago; today they share 98.4% of their DNA. The characters distinguishing man from chimpanzee, like the upright posture, a larger brain, the ability to speak, are therefore only due to that 1.6% of difference in their DNA, or better only to an unknown fraction of that 1.6% because some differences in the proteins have been identified and shown to play no role in the determination of the distinctive characters.

Chimpanzees are then our closest 'cousins' (the difference from gorillas is barely larger at 2.3%), so close in fact that according to Jared Diamond (*The Third Chimpanzee*), it can be stated that today on the earth live three species of the genus homo, the common chimpanzee (homo troglodytes), the pygmy chimpanzee (homo paniscus) and the third chimpanzee (homo sapiens).

Female bonobo with infant at San Diego Zoo

From fossil skeletons, it has been deduced that part of this 1.6% (or fraction) of genetic difference had already been covered by man about 40,000 years ago, when the brain achieved 'modern' size, the genetic difference between us and that man is therefore much less than 1.6%. And yet this 'small' difference is responsible for the sudden acceleration in the evolution of man, characterised by extremely slow changes for millions of years and extremely fast in the last 40,000 years.

'What', in this small genetic difference, may have produced such enormous consequences?

One possibility, as suggested also by Diamond (cit.), is that at some point in the evolution, a mutation in one of the hominoids produced an alteration of the vocal tract (larynx, tongue and associated muscles) such as to allow the formation of a greater variety of sounds as compared to our cousins, providing in time the anatomical basis for the birth of a complex language which, in turn, produced writing.

These two elements are certainly of fundamental importance in the development of modern man, and if they were due to a mutation, a new piece of improbability must be added to the existence of homo sapiens (remember that mutations are entirely casual).

A new difference has been recently discovered in man's cerebral cortex, the presence of some genes and a particular kind of neurons which do not exist in our cousins, chimpanzee, bonobo, and gorilla. These neurons are rare, less than 1%, but they can regulate typical human functions, like memory and behaviour, and diseases like Parkinson's or some forms of dementia [AMM Sousa *et al, Science 2017*]. These different genes are also a consequence of mutations.

The probability of occurrence of two independent events (independent means that the occurrence of one is in no way influenced or determined by the occurrence of the other) is given by the product of the probabilities of the two single events. Then the probability that, given the initial conditions on the new-born earth, 'appears' homo sapiens is the product of many probabilities, all quite small, like:

– birth of the first replicant
– birth of the first prokaryotic cell

– birth of the first eukaryotic cell

– extinction of the Ediacara fauna

– extinction of the Burgess fauna and survival of Pikaia

– Cambrian extinction

– Permian extinction

– mutations determining human characteristics, like the birth of language.

We do not know the probabilities of these single events, but just for an example, let us assume that each of them has a probability to occur of 1 out of 10. Then the probability of the appearance of homo sapiens is the product 8 times of 1/10, that is 1 over 100 million: this means that starting again from the beginning (may be on some other planet), evolution could follow 100 million different paths, only one of them leading to homo sapiens.

And this is an obvious over-estimate since the probability of each of the events considered is certainly much smaller than 1 out of 10. The conclusion is that the probability of the appearance of homo sapiens is negligibly small.

5. The Universe

It is a rather embarrassing situation to have to admit that we are unable to find 90% [of the matter] of the Universe [Bruce H Margon, *New York Times*, 2001]

Up to some years ago, we thought we understood many of the mysteries of the universe, to know how it was born about 13.7 billion years ago, how it evolved to form its present content of stars—grouped into galaxies—dust and radiation of all frequencies. Not anymore. Today, it is believed that all these 'ordinary' components constitute only a small percentage of the universe, the rest is made of quantities, still unknown, named 'dark matter' and 'dark energy'. To understand what these 'dark things' are, since we do not know, but how and why the hypothesis of their existence arose, let us look at the universe that we know.

Before discussing 'what' we know about the universe, it is important to understand 'how' we know it, the methods and tools of investigation.

Some of our information comes from the study of particles reaching the earth, cosmic rays from all space, the solar wind from the solar surface, neutrinos from the solar interior and from the explosions of supernovae. Information is also found from the analysis of materials collected in space missions to the moon, Mars, and comets or from meteorites which reach the earth.

Apart for these exceptions, all we know—or do not know—about the universe is deduced from the observation of the electromagnetic radiation ('photons') emitted or absorbed by all objects in space. The whole electromagnetic spectrum, that is radiations of all frequencies, is present in the universe; different objects and different physical processes are responsible for the emission or absorption of radiation at different frequencies, corresponding to photons of different energies. The knowledge of the physical laws governing emission, absorption and propagation of radiation in its interaction with matter, allows to deduce, from the observation of the spectrum, the physical conditions of the emitting or absorbing objects and of the matter through which the radiation passed.

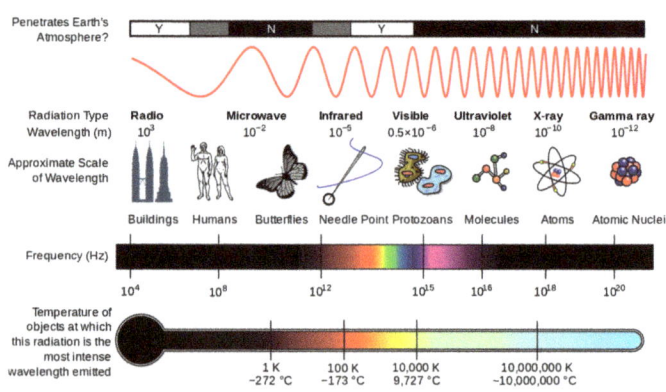

The Electromagnetic Spectrum

A very important kind of radiation is the so called 'black body' radiation: electromagnetic radiation of all frequencies emitted by any body, whatever its temperature, with the intensity of the emitted radiation at the different frequencies depending—with a known law—on the body temperature. The temperature of a body can then be deduced from a measure of the frequency corresponding to the maximum intensity of the emitted radiation. For instance, incandescent bodies, like a heated iron bar, emit radiation with maximum intensity at the frequencies (colours) to which our eyes are sensible, the 'visible' part of the spectrum.

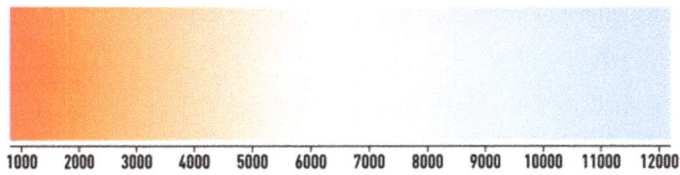

Colour of a black body from 800 K to 12200 K. This range of colours approximates the range of colours of stars of different temperatures, as seen or photographed in the night sky.

The 'visible spectrum' is one of the frequency (or wavelength) bands into which the electromagnetic spectrum is conventionally divided; it is defined by the limits of the human eye, while other parts of the spectrum invisible to the human eye, like radio-waves, infrared, ultraviolet, X-or gamma-rays, with longer or shorter wavelengths (frequencies) can only be 'seen' with appropriate detectors. The term 'light' usually refers only to the visible part of the spectrum, this is the part which can pass through the earth's atmosphere, and this is probably the reason why the human eye evolved with sensitivity to this part of the spectrum. Our atmosphere is transparent to radio-waves as well but the intensity of radio-waves reaching Earth is negligible as compared to the intensity of light, coming mainly from the sun. All other frequencies in the spectrum are absorbed by the atmosphere and to detect them it is necessary to have telescopes above the atmosphere.

That is why astronomy, until 1945, has been an 'optical' astronomy, observing the sky with ground-based 'optical' telescopes which can only see a very limited part of the universe and the image of the universe was then the one that a very powerful human eye (eye+lenses) could see. This image started to change when the first radio-telescopes were built, which could 'see' objects emitting radiation in the radio-frequency band, and changed again when the first space-telescopes, like Hubble and Spitzer, started to observe the sky from their orbits above the atmosphere and, using new detectors developed to measure radiation at all frequencies, started to explore the sky in frequency bands previously invisible. Today, we 'see' the universe in the full electromagnetic spectrum, from the very low-frequency infrared radiation to the extremely high frequency of the gamma rays.

This is the continuous spectrum, the emission is at all frequencies and is due to the motion of electric charges: any object radiates because atoms and molecules of any object are always in translational and/or vibrational motion and their kinetic energy is transformed into the energy of the emitted electromagnetic waves, radio waves are generated by the motion of very fast electrons in an ionised gas, X-rays are emitted by relativistic electrons moving in a magnetic field.

Since the emission of radiation at different frequencies is due to different objects and physical processes, the universe we see in the various frequency bands is quite different. For instance, 'cold' bodies emit in the infrared band and if we explore the sky with an infrared space telescope, we see mainly the large, cold clouds of gas and dust, the birth-site of stars; if we look through an X-ray space telescope, we mainly see those regions where the physical conditions are consistent with the generation of high energy electrons which emit in the X-ray band.

The following images show different objects as seen by space telescopes operating in the X-ray and infrared bands. The NASA space telescope Chandra explores the sky in the X-ray band.

X-ray image of the sun, the emission (shown in green and blue) originates from regions with temperatures higher than three million degrees.

Infrared image of the gas cloud W40 in our galaxy.
The birth of some stars can be noticed.

Image in the red, blue and infrared bands of the surface of Pluto

The expanding residue of a supernova exploded in one of the
Magellanic Clouds, a nearby galaxy, as observed by Chandra in
the X-ray band (blue in the image) and by Hubble in the visible
band (red in the image).

The sky in the X-ray band as seen by Chandra. The image is the result of seven million seconds of observation, the red, green and blue colours represent emission of X-rays of low, medium and high energy respectively.

The continuous spectrum of radiation is not the only one to provide information on the nature of the universe, more come from the 'discrete' spectra or 'line' spectra. These include the emission and absorption spectrum, fundamental tools to identify the elements which are present in the sources of radiation, it is from the analysis of these spectra that the chemical composition of the universe was discovered and found to be made of the same chemical elements known on the earth.

A look at the discrete spectrum is appropriate. When the electrons in an atom make a transition from a higher energy to a lower energy state, electromagnetic radiation is emitted with a frequency (energy of the photons) determined by the energy difference between the two states. Many such transitions are possible for each chemical element or compound and the 'emission spectrum' of an element or compound is defined as the ensemble of all the frequencies of the radiation emitted by that element or compound in the transitions between its energy states. In the inverse process, an atom can absorb electromagnetic radiation and use its energy to 'force' its electrons to make transitions from states of lower to states of higher energy and, as in the emission process, the energy difference between the two states is determined by the frequency of the absorbed radiation.

The 'absorption spectrum' of an element or compound is defined as the ensemble of all the frequencies of the radiation that the element or compound can absorb. Since the possible transitions between states, one way or the other, are the same every element or compound can emit and absorb the same frequencies and, most important, these frequencies are typical for each atom or compound, like fingerprints, so that the emission and absorption spectrum of each atom or molecule is unique, each spectrum belongs to one and only one atom or molecule. The analysis of these spectra, called 'spectroscopy', provides therefore information on the chemical composition of the observed object.

Unlike the continuous spectrum, which contains all frequencies (all the colours in the visible part), the absorption and emission spectra contain only 'lines', corresponding to the particular frequencies (colours in the visible part) emitted or absorbed.

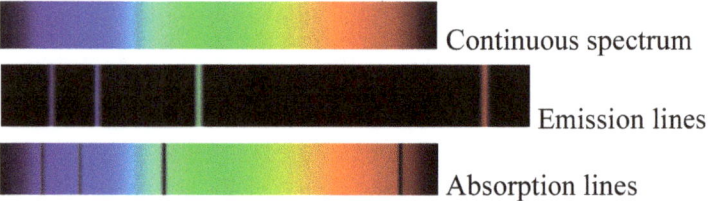

Continuous spectrum

Emission lines

Absorption lines

The number of lines in the spectrum of an element depend on the number of possible transitions between states, that is the number of electrons in the atom of the element.

Emission Spectrum of Hydrogen in the visible band

That is why the iron atom, with 26 electrons, has a spectrum with many more lines than the hydrogen atom, which has just one electron.

Emission Spectrum of Iron in the visible band

The continuous and discrete spectra are our basic tools to explore and study the universe. With the large optical telescopes, we can see about 100 billion galaxies, each formed by about 100 billion stars, large dust clouds and interstellar gas. The number of galaxies in the universe is certainly much larger, a recent estimate (2016) is 2000 billion, including the ones we cannot observe.

Since the electromagnetic radiation travels with a velocity large but not infinite (about 300,000 km/sec), it takes time for the radiation emitted by an object to reach our telescopes, we 'see' an object not as it is at the moment we observe it, but as it was when it emitted the radiation. For instance, it takes eight minutes for the solar light to reach us, that is when we look at the sun, we see it as it was eight minutes before. And we 'see' a galaxy as it was millions or billions of years ago.

A convenient unit to express astronomical distances is the light-year (ly), the distance covered by light in one year, about 9000 billion kilometres. In these units, the distance of the sun is eight light-minutes, the closest star, Proxima Centauri, is 4 ly away, our galaxy, the Milky Way, has a diameter of 100.000 ly, the closest galaxy, Andromeda, is 2.5 million ly away and the farthest observed galaxies are billions of light-years away. When we look at the universe, we see images of a far past.

The light reaching us today from the farthest observable galaxies was emitted billions of years ago, the light from even farther galaxies will never reach us because the universe is expanding.

The boundary of the universe, that is the radius of the observable universe, would be at the distance travelled by light in the time since the birth of the universe, that is at 13.7 billion light-years, if the universe had not been continuously expanding. But since the universe is expanding, the light which left a source, a galaxy for instance, 13.7 billion years ago, belongs to a source which in the meantime moved away from us due to the expansion. The boundary of the universe therefore keeps moving away, a recent estimate puts it at more than 45 billion light-years. We will never observe it.

The Stars

From the analysis of what we can observe, combined with the laws of physics and chemistry, we reached an understanding of the nature, structure and history of the fundamental objects which constitute (or so it was thought) the universe: the stars. These differ from each other by the amount of energy emitted, called the luminosity, and for their surface temperature (the colour).

To understand how and why the stars evolve from birth to death, we need to understand first why the stars shine.

At the start of last century, the source of the enormous amount of energy emitted by stars for billions of years was still a mystery. All the solutions proposed led to estimates of the sun's age smaller than the earth's, estimated at 4.5 billion years. Take chemical energy, for instance, like combustion, that is the combination of carbon with oxygen. From the knowledge of the energy produced in combustion and the amount of energy emitted by the sun every second, we can calculate how long the sun would live emitting the energy produced by combustion. The result, in the most favourable hypothesis that the sun's composition is half carbon and half oxygen, is 5400 years! A completely different source is needed to make the sun (and the stars) shine for billions of years.

The solution came from new discoveries in nuclear physics in the 1930s, when it was understood that at temperatures of millions degrees, as in the solar interior, the atomic nuclei 'fuse' to form heavier nuclei whose mass is not exactly the sum of the masses of the original nuclei and the mass difference is transformed into energy according to Einstein's equivalence principle of mass and energy (these processes are called thermonuclear reactions). For hydrogen atoms, the main component of the sun, four hydrogen nuclei fuse to form a nucleus of helium and releasing part of the energy as electromagnetic radiation, high energy (frequency) X-and gamma rays. The enormous number of such reactions in the solar interior produces the energy needed to balance the gravitational collapse and keep our star shining.

The high energy photons generated by the thermonuclear reactions in the deep solar interior, where the temperature is sufficient to trigger these reactions, travel towards the surface interacting with matter through 'diffusion' processes of absorption and re-emission at lower energies, and finally emerge from the solar surface with much lower energies, that is with frequencies in the visible part of the spectrum. The result is a present for us, the solar light.

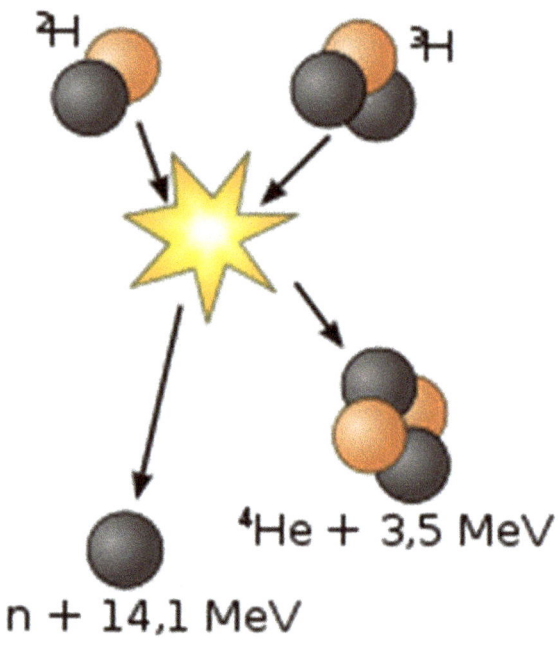

^{4}He + 3,5 MeV

n + 14,1 MeV

The fusion of Hydrogen, the energy released is in million electronVolt (MeV).

To grasp the incredible efficiency of this energetic source, let us consider some numbers. Every second in the solar deep interior 600,000,000 tons of hydrogen fuse to form 595,740,000 tons of helium and the mass difference, 4,260,000 tons of hydrogen are transformed into energy (E) according to Einstein's equation $E=mc^2$. It may look like a huge mass loss but, with this rate, the sun loses only 0.06 per thousand of its mass every billion years.

The energy so generated *every second* is equivalent to 106 400,000,000-terawatt hour (TWh), the world production of electric energy is about 25.000 TWh in *one year*: the sun produces in one second the energy that all the production plants of electric energy of our planet would produce in four million years.

Thermonuclear fusion is a clean source of energy, unlike nuclear fission, the energy source used in nuclear reactors (and the atomic bomb), there is no production of radioactive scores. That is why we have been trying for years to imitate the stars and reproduce thermonuclear fusion in the laboratory, this solution to our energetic problem would also have the advantage of the enormous amount of available fuel, just think of how much hydrogen there is in the oceans!

The fusion of hydrogen is the energy source of the stars born inside molecular clouds, constituted mainly by hydrogen, when a local increase of the cloud density, due for instance to the shock waves coming from the explosion of a nearby star, induces the gravitational attraction of surrounding material and the whole cloud starts to collapse due to its own gravity, forming a continuously growing mass around the 'initial' density lump. During the collapse the gravitational energy of the particles 'falling' towards the interior decreases while their kinetic energy (velocity) increases, exactly as it happens for any object falling to the ground from some height. The kinetic energy of an ensemble of particles is the temperature of the ensemble, so the temperature of this 'proto-star' keeps increasing during its collapse, until it reaches (in the deep interior) the value needed to trigger the fusion of hydrogen into helium. At this point, the energy produced stops the collapse, the proto-star starts to shine, a star is born.

This process however can only take place if the proto-star reaches a mass of at least 0.08 solar masses, for lower masses, the temperature cannot reach the value needed to trigger the fusion of hydrogen and the proto-star reaches an equilibrium as a cold, low-brightness 'brown dwarf'.

The stars which live transforming hydrogen into helium in their interior are stable, with constant pressure and temperature, and their representative points are found to occupy a particular region in a colour-luminosity diagram (the H-R diagram), called the Main Sequence. The sun is a main sequence star and will be such for about ten billion years. Stars brighter and more massive than the sun exhaust their hydrogen 'fuel' faster and their permanence on the main sequence is shorter (tens to hundreds million years), while stars less bright and massive than the sun 'burn' their fuel more slowly and can stay on the main sequence for up to hundreds of billion years.

The main sequence in the H-R diagram contains therefore stars of all masses, age and brightness (ranging in colour from red to blue, with the yellow sun in the middle), with the common feature of transforming hydrogen into helium in their core.

But sooner or later, the moment arrives for all when the hydrogen in the core has been completely transformed into helium by the thermonuclear reactions. The star has now a structure formed by a core of helium surrounded by a layer of 'unburned' hydrogen, and, without the energy source, the gravitational collapse starts again and the internal temperature starts to rise again at the expenses of gravitational energy.

The successive evolution of the star—and the movement away from the main sequence of its representative point on

138

the H-R diagram—depend on the mass of the star since the mass determines the amount of gravitational energy available to increase the temperature. Stars with mass of the order of the solar mass can reach central temperatures as large as 100 million degrees, sufficient to trigger the thermonuclear fusion of the helium nuclei into carbon and oxygen releasing energy again. Stars with smaller masses never reach the temperature for the burning of helium, their collapse continues until the central density becomes so high that the electrons form a 'degenerate gas' (a quantum effect) whose pressure balances the gravitational fall and they reach an equilibrium as low luminosity stars, formed essentially of helium, known as 'white dwarfs', Their representative points move away from the main sequence of the H-R diagram.

This is not the place to follow the details of stellar evolution after leaving the main sequence, to the interested reader I can suggest one of the texts treating stellar evolution at the popular level (no mathematics!), Longair's *The New Astrophysics* in Davies' book *The New Physics.*

Here, to continue our story, it is sufficient to mention some phases of the post-sequence stellar evolution, in particular the origin of chemical elements and black holes.

Post-sequence evolution is determined by the combined role of gravitational collapse and fusion—or non-fusion—of the elements in the stellar core and/or in external layers, like the hydrogen layer surrounding the helium core or in the following phase, the helium layer surrounding the carbon/oxygen core after the fusion of helium. Mass loss also plays an important role. All stars lose mass ejecting particles from their surface, the so-called 'stellar winds'. We have a direct experience of the wind from our star, the solar wind, made mainly of electrons and protons emitted from the solar surface, continuously flowing around the earth (and all the planets in the solar system), causing phenomena like the auroras and disruptions in our communications. And in some evolutionary phases, the stars can 'expel' their external layers, which in some cases form a cloud surrounding the stellar residue (these objects are known as 'planetary nebulae') and in other cases, where the expulsion of the layers is explosive, the resulting object is known as a 'nova' or 'supernova'. Given the fundamental role mass plays in the temperature reached by the stellar core, that is triggering or non-triggering of the fusion reactions, the amount of mass lost in these processes can play a key role in the evolutionary path of a star, or of whatever is left after the ejection of mass.

As we have seen, at the end of the main sequence phase a star has a core of helium surrounded by a layer of hydrogen. If the energy released by the gravitational energy is sufficient to reach the temperature for the fusion of hydrogen in the layer, the energy released in these reactions causes the expansion of the superficial layers of the star which, upon expansion, cool down: the star radius increases, its surface temperature decreases, the colour tends towards the red. These stars are known as 'red giants'.

This is the next step of the sun, which today is half-way through his life as a main sequence star. At the end of this stable period, in about five billion years, the sun will evolve into a red giant, expanding beyond Mercury's orbit.

At the end of the main sequence phase, some stars live burning their helium core. What happens next? If their mass is sufficient, through a sequence of collapses they can reach the temperature needed to trigger the fusion reactions of carbon into heavier elements, like Neon, Magnesium, Silicon and Sulphur.

If there is enough mass to provide the energy to reach the temperatures needed for the next fusion reactions, also these elements will 'burn' and the nucleosynthesis can proceed towards heavier elements. Up to Nichel-56 (^{56}Ni) which decays into iron-56 (^{56}Fe): this is the most stable element (maximum binding energy) and nucleosynthesis cannot proceed farther. Now, the star structure resembles an onion, with hydrogen in the most external layer, followed by a helium layer, then a carbon layer and so on with layers of heavier and heavier elements, produced in the successive fusion reactions.

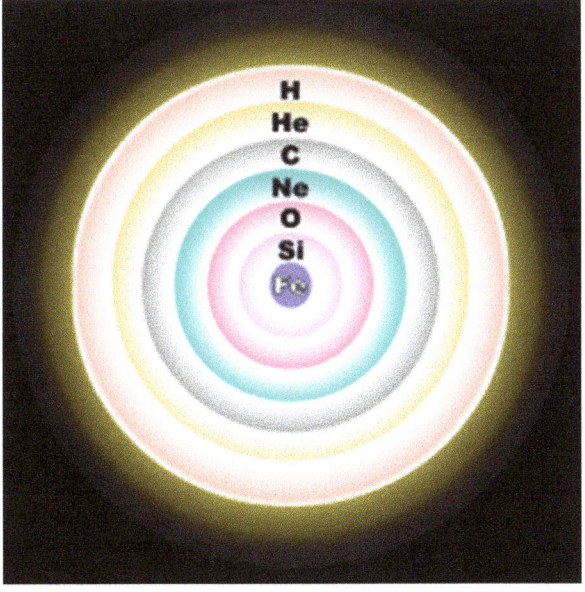

Schematic representation of the 'onion structure' of a massive star in a late evolutionary state (Not to scale)

When a star ejects its external layers or explodes like a supernova, the elements which were 'built' in its interior are freed in the interstellar medium. Stars are the 'cosmic factories' of all chemical elements heavier than helium (up to iron) present in the universe, while hydrogen and a part of helium have been produced, as discussed later, shortly after the birth of the universe before stars were formed.

Except for hydrogen, all the atoms which constitute our bodies have been 'manufactured' billions of years ago in some star somewhere in the universe.

Elements heavier than iron have their origin in the explosions of supernovae. These events are produced in stars with mass larger than eight solar masses when energy production in their core is reduced to the point that it can no longer support its own mass and its sudden gravitational collapse induces a catastrophic explosion of the whole star into an extremely bright supernova. The luminosity of the supernovae is huge, it can even be greater, for a few days, than the luminosity of the whole galaxy to which they belong. Those exploded in our galaxy in historical times, like the one exploded in 1054, whose residue we observe today as the 'Crab nebula', were visible to the naked eye and considered as new stars, whence the name 'nova' (new).

The temperature reached in a supernova explosion is such that there is enough energy for the products of the stellar nucleosynthesis to 'fuse' into heavier elements such as Gold, Magnesium, etc.

The external layers ejected in a supernova explosion diffuse in the interstellar medium as a very hot gas, expanding to form a bright 'nebula', the spectacular Crab nebula is shown in the figure. While the external layers travel outwards, the residual star core 'implodes' and its density increases to such high values that the electrons are 'captured' by the protons and the electron-proton gas in the core is transformed into a gas of neutrons. What happens next depends on the mass of the residue. If this is between 1.4 and 3.8 solar masses, the pressure of the degenerate neutron gas (the same quantum effect we have seen for the degenerate electron gas in white dwarf stars) can stop the collapse of the core and the residue reaches a stable equilibrium as a 'neutron star'.

The Crab Nebula, residue of a supernova exploded in 1054 (photo by the space telescope Hubble).

Neutron stars are not directly observable, but their existence has been proved by the discovery of 'pulsars', objects emitting a periodic radio signal due to their rotation, like a lighthouse, shown to be fast rotating neutron stars. One has been detected at the centre of the Crab Nebula, the end product of the stellar core in a supernova explosion, while the expanding external layers of the initial star form now the bright visible nebula.

For stars so massive that the residual core after the explosion has a mass larger than 3.8 solar masses, the pressure of the degenerate neutrons is not sufficient to stop the gravitational collapse, the core radius keeps decreasing and density increasing until the gravitational field of this 'object' becomes so intense that Newton's classical theory is no longer valid, the description of such 'objects' requires Einstein's theory of gravity, General Relativity. Using Einstein's theory, it has been found that the gravitational field of these objects can induce a curvature of space-time around the object such that not only matter but even radiation cannot escape, these objects cannot be 'seen', they have been named 'black holes'.

Their existence, as for neutron stars, is a theoretical prediction, a direct consequence of the theory of stellar structure and evolution.

And, as for neutron stars, also the existence of black holes has been proved by the observation of their effects. Many stars form 'binary systems', two stars orbiting around each other: if one of the two evolves into a black hole, its intense gravitational field will produce tidal effects on its companion star which can be so strong as to tear it apart into a stream of gas which is then 'swallowed' by the black hole. The theoretical modelling of this process shows that initially the 'falling' matter from the destroyed star forms a disk orbiting around the black hole and, when new matter 'falls' onto this disk, the interaction produces emission of intense radiation, especially in the X-ray band. It was assumed that this radiation could be observed since it originates outside the black hole. And indeed, it has been observed. In fact, many cosmic X-ray sources have been observed with characteristics of the radiation consistent with the theoretical models. In 2014, for the first time, even the 'birth' of an X-ray source was observed, a star in a binary system 'sucked' by its black hole companion 290 million years ago.

An artist's drawing of a black hole named Cygnus X-1. It formed when a large star caved in. This black hole pulls matter from the blue star beside it.
Image: M Weiss/NASA/CXC

Artistic view of the event observed by the NASA SWIFT telescope: material from a star shredded by the gravitational field of a black hole forms an accretion disk. The event has been observed thanks to the radiation emitted by the disk in the X, ultraviolet and optical bands.

A similar event was observed in 2019 by the NASA mission TESS, while searching for exoplanets, as discussed in the next chapter.

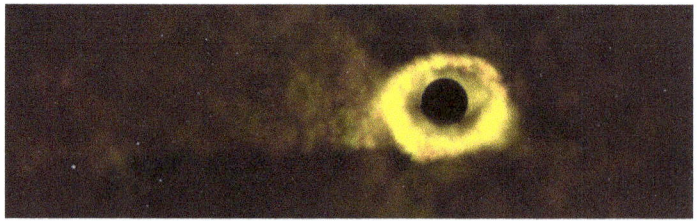

When a star strays too close to a black hole, intense tides break it apart into a stream of gas. This video includes images of a tidal disruption event called ASASSN-19bt taken by NASA's Transiting Exoplanet Survey Satellite (TESS) and Swift missions. Credits: NASA's Goddard Space Flight Centre

And in 2019, the international network of radio-telescopes EHT (Event Horizon Telescope) captured the first ever 'image' of a super-massive black hole (or better of its shadow) in the M87 galaxy, about 55 million light years away: the black hole, of course, is…black, but it appears to cast a shadow against the bright background of the hot disk of material surrounding it and the telescope could see the light 'bending' in its intense gravitational field as Einstein's theory predicts.

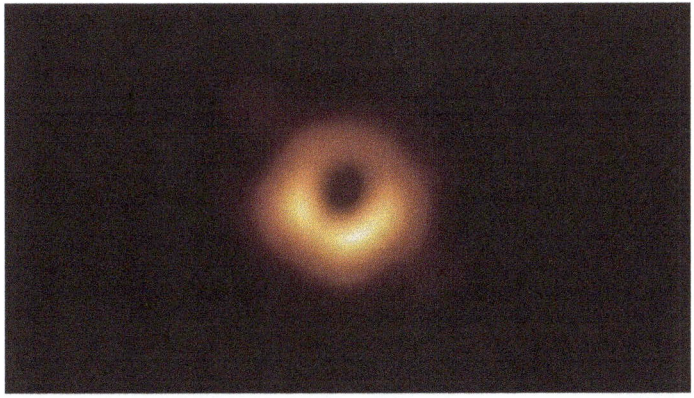

First image of the shadow of a black hole at the centre of the galaxy M87. The image shows a bright ring formed as light bends in the intense gravity around a black hole that is 6.5 billion times more massive than the sun. Credit: Event Horizon Telescope Collaboration

The discovery of pulsars and binary X-ray sources is a confirmation that the theory of stellar evolution is correct all the way to the final phases of a star's life, neutron stars and black holes, the 'death' channels of stars which started their life on the main sequence burning hydrogen in their core.

Galaxies: The Expanding Universe

Stars are born, live and die inside galaxies, which in turn are organised into groups, clusters (up to 1000 galaxies) and superclusters. Our galaxy, the Milky Way, is part of a group, called 'the local group', formed by two clusters of galaxies whose largest members are our galaxy on one hand and the Andromeda galaxy on the other. They are both spiral galaxies and each have their own system of smaller 'satellite' galaxies, like the Large and Small Magellanic clouds (satellites of the Andromeda galaxy). The local group is part of the larger Virgo supercluster but the total number of galaxies in the group is not known, it is certainly in excess of 54, most of them dwarf galaxies.

The Spiral Galaxy M101, 21 million light years away

Our knowledge, as well as our ignorance, of the large-scale structure of the universe and of its origin stems from the observation of galaxies.

Our present picture of the universe derives from the combination of the spectroscopic observations of galaxies with Einstein's General Relativity. The first show that the spectral lines of stars in clusters of galaxies sufficiently far away (a few millions light years) are red shifted.

The red-shift or blue-shift of the radiation is known as the Doppler effect: the electromagnetic radiation emitted by a source moving with respect to an observer is found to have a wavelength larger (if source and observer move away from each other) or smaller (if source and observer move towards each other) than the wavelength of the radiation as emitted from the source. If the radiation is visible light, the effect is a shift towards the red part of the spectrum in the first case, towards the blue in the second case.

For nearby galaxies, like the ones in the local group, their gravitational attraction pulls them closer and blue-shifted spectral lines are also observed. But for galaxies so far apart that their gravitational interaction is negligible, the spectroscopic data show a red shift without ambiguity, meaning that these galaxies are moving away from each other: the universe is expanding.

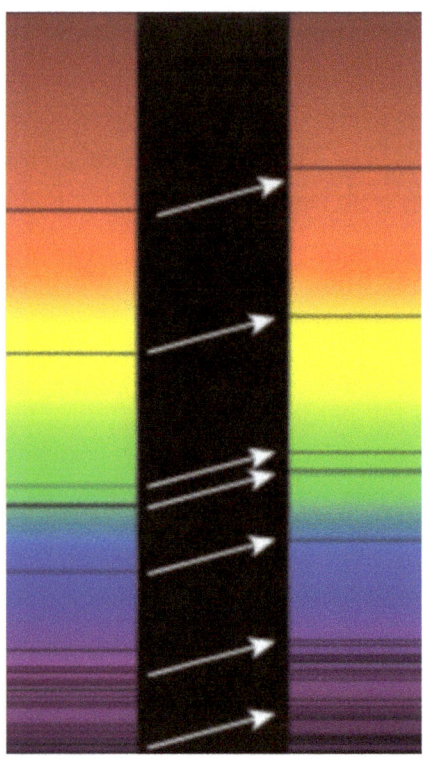

Lines in the optical spectrum of a cluster of far galaxies (right) compared to the absorption lines (left) of the sun's optical spectrum. The wavelength increases (the frequency decreases) as the galactic lines are shifted towards the red, as shown by the arrows.

An expanding universe is consistent with mathematical models of the universe proposed as long ago as 1922 by the Russian cosmologist Alexander Friedman, from the solution of the equations of general relativity, in contrast with the solution proposed by Einstein, which is consistent with a static universe. Friedman could show that the only solution of

Einstein's equations compatible with the assumptions of homogeneity and isotropy of the universe (meaning that the universe looks the same for all observers, whatever their position, there are no 'privileged' points of observation) is the one where the universe is expanding and depending on the average density of matter in the universe, there are two possible modes for the expansion. If the average density of matter is lower than some 'critical density', the gravitational attraction between galaxies will slow the expansion but will not stop it (open universe), if larger than the critical density it will eventually stop it and a gravitational contraction of the universe will then follow the expansion (closed universe).

It is also possible to imagine an 'oscillating universe' where the cycle of expansion-contraction repeats indefinitely: at the end of each contraction, an expansion sets in, it slows down till it stops and a new contraction begins...

A cluster of galaxies photographed by the Hubble Space telescope. They all appear more reddish due to the expansion and, from the measure of the redshift it can be deduced that the light we see today left the cluster more than three billion years ago. (The bright star-like objects are nearby stars which happened to be in the field-view of the cluster).

The red shift of far galaxies and Friedman's mathematical model form the basis of our present view of the expanding universe, not as matter expanding in empty space but rather as an expansion of space itself. There are two frequently quoted analogies to 'represent' this expansion. In one of them, the universe is a raw cake (representing space) with raisins (representing galaxies): in the oven the cake expands and the distances between the raisins increase, *even though each raisin keeps its place in the cake.* In the other one, the universe is a balloon (representing space) with ink dots (representing galaxies) on its surface: when the balloon is inflated the distances between the dots increase, *even though each dot keeps its place on the balloon.*

The Big Bang

If the galaxies are moving away from each other, we can start from the present situation and reverse the arrow of time, tracing the evolution of the universe back in time, when the galaxies were closer and closer...Doing so, we reach an extremely dense and hot state in the past and the theory which describes how the universe expanded from this initial state, and is still expanding today, is called the Big Bang theory, or the Standard Model, even though the theory does not assume any explosion, the name was actually coined by the astrophysicist Fred Hoyle, who did not use it, in a derogatory sense.

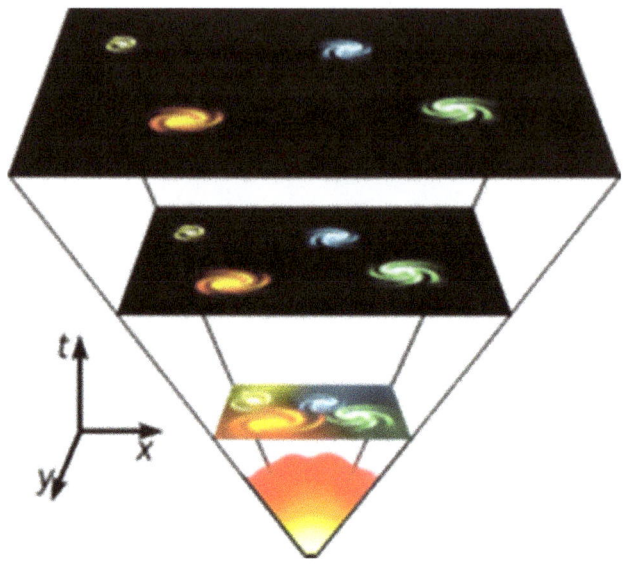

The expanding universe: reversing the time arrow reach an initial 'singularity', the Big Bang.

With the words of the astrophysicist PJ Peebles:

The essence of the Big Bang is that the Universe is expanding and cooling. Notice that I said nothing about an 'explosion', the Big Bang theory describes how the Universe evolves, not how it began.

With no reference to bangs or explosions, the name Big Bang indicates the dense and hot state that 13.7 billion years ago was the 'birth' of the universe, defined as the beginning of its expansion, while the theory is not concerned with what happened 'before' the Big Bang: *asking what happened before the Big Bang is like asking what there is north of the North pole* [S Hawking in P Davies *The New Physics*].

May be there was a 'beginning' or maybe not, if we follow the reversed arrow of time all the way to 'time zero', eventually a point of infinite density will be reached, called a singularity, where the laws of physics, even the concepts of space and time, have no meaning. Clearly, we have no means to describe such singular state, if it ever occurred it must be left in the unknown region 'before' the Big Bang, which instead must start at a time when the state of the universe was very hot and dense but such that the laws of physics can describe it.

The choice of the initial state is arbitrary to some extent and is up to us. The choice was to 'start' the universe one hundredth of a second after its 'birth' (whatever that was), when the temperature was already 'down' to 100,000 million degrees Kelvin from the much higher values it had in the first hundredth of a second. This choice was motivated by the fact that this particular temperature is the threshold temperature for the creation of the electron and its anti-particle, the positron. At higher temperatures, according to Einstein's equivalence of mass and energy, hadrons, like pi-mesons and others, can be generated from the radiation field if its energy (temperature) is larger than the rest mass energy (mc^2 of the particles. Hadrons interact via the 'strong force', and when the Standard Model was formulated, the necessary theory to describe their properties was not available.

With the choice of this threshold temperature therefore it was possible to assume that in the 'initial' state no hadrons are present, only electrons and positrons, whose interactions are well known, together with the massless particles, neutrino and antineutrino.

Thus, one hundredth of a second after its birth, the universe was a 'soup' of electrons, neutrinos (with their antiparticles), a few baryons (protons and neutrons) and of course photons (the radiation field), all in thermal equilibrium at the same temperature of 100,000 million degrees Kelvin. This is the Big Bang.

From the physical characteristics of this initial state, the theory follows the evolution of the universe as it expands and cools. Its present success is in the fact that, unlike other theories, the Big Bang theory leads to 'observable' previsions confirmed by observation.

In the initial phases, the temperature was so high that the energy of the photons disrupted any bonds between electrons and protons so that atoms could not be formed. As we have seen, when radiation and matter are in equilibrium at the same temperature, the radiation has a 'black body' spectrum, therefore the Big Bang was characterised by black body radiation at a temperature of billions of degrees: after expanding and cooling, the universe must still be permeated by this 'fossil' radiation, of course at a lower temperature. The theory can calculate what should be the temperature and the spectrum of this radiation after about 13 billion years, and the result is: a black body spectrum at a temperature of about 3K (Kelvin degrees).

The most important confirmation is the existence of an electromagnetic radiation, called Cosmic Microwave Background Radiation (CMBR), permeating the whole universe, it is not associated to stars or galaxies or any other cosmic body and has maximum intensity in the microwave region of the electromagnetic spectrum. The CMBR, discovered in 1964 by the American astronomers Arno Penzias and Robert Wilson, is considered the residual radiation from the initial phases of evolution of the universe, consistently with the results of the Big Bang model.

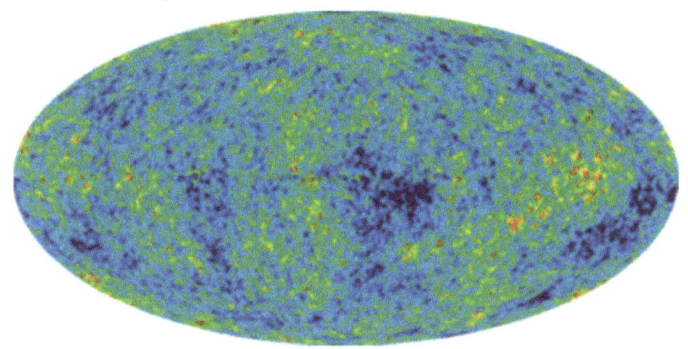

Nine-year WMAP image (2012) of the CMBR across the universe.

In 1990, the NASA COBE satellite (*COsmic Background Explorer*) found that the CMBR has indeed a thermal black body spectrum at a temperature of 2,725 K, almost uniform in all directions, meaning it does not come from any direction but 'fills' the entire universe. No other model of the universe, so far, has been able to explain the existence of the CMBR.

The other great success of the Big Bang theory is the cosmic abundance of the light atoms, hydrogen, deuterium and helium. This seems to be today, and have been in the past, the same everywhere in the universe and is in excellent agreement with the amounts predicted by the mechanisms of formation of these elements in the Big Bang model.

The Big Bang temperature started to decrease in the expansion and when, a few minutes after the initial state, it reached about one billion degrees Kelvin, the neutrons could combine with the protons to form the first nuclei of the light atoms (Deuterium is made by one proton and one neutron, helium by two protons and two neutrons), but the major part of protons remained free, as nuclei of hydrogen, and this explains why hydrogen is the most abundant element in the universe. The temperature kept on decreasing and when it reached thousands of degrees Kelvin, the energy of the photons was no longer sufficient to avoid the formation of the first stable atoms from the combination of the free electrons with the hydrogen, deuterium and helium nuclei. The light elements were thus formed, and their relative abundance has since then remained the same everywhere in the universe, apart for small variations due to their consumption and/or creation in stellar interiors, where the elements heavier than helium originate.

At this stage, the photons could 'detach' themselves from the electrically neutral atoms and freely travel through space, leading to a full 'decoupling' of matter and radiation. Since then the photon temperature has steadily decreased, it is presently 2,725 K, and keeps decreasing as the universe keeps expanding.

The existence and properties of the Cosmic Microwave Background Radiation and the observed abundances of the light elements constitute the fundamental proof of the Big Bang theory.

But there was a problem that the theory could not explain: the observed uniformity of the Cosmic Microwave Background and of the universe itself on a large scale. The Standard Model must 'assume' that the universe was uniform from the very beginning but then it cannot explain why it is

158

not uniform on a small scale, where it is filled with lump-like galaxies and clusters of galaxies.

A model known as 'inflationary universe' was then proposed, initially by Alan H Guth in 1980, to solve these problems. It looks at what happened before the Big Bang, assuming that the universe was much smaller, so small that matter and radiation where in such close contact that uniformity dominated. Then in an extremely small fraction of a second (10 to the power -32 seconds), an accelerated expansion took place increasing the radius of this tiny initial bubble of an enormous amount (10 to the power 25). This accelerated expansion is called 'inflation' and is caused, according to this theory, by the energy of the vacuum of the quantum fields. We have seen before that the various fields that permeate empty space have a vacuum energy; the theory assumes that a sudden fluctuation in the vacuum energy density of the fields could drive such explosive expansion.

The British cosmologist Sir Roger Penrose, Nobel laureate, has even proposed that the universe was born because of a sudden fluctuation in the field energy of empty space. In the beginning, there was vacuum…

It should be mentioned that an expanding universe with an origin in time also explains the famous paradox proposed by Olbers in 1826. If there are infinite stars homogeneously distributed in the universe, looking in any direction, I should see the light of a star and the whole sky should therefore be as luminous as a stellar surface: why then the nocturnal sky is black?

If the universe exists from a finite time, the light from far stars did not reach us yet and, if the universe is expanding, the light from far stars will never reach us. In both cases, the light reaching us comes from a limited number of stars, such that the sky appears to be black.

Dark Matter and Dark Energy

The Big Bang theory seems then to provide the correct explanation for many of the things we observe in the expanding universe, but it cannot explain the most recent observational data on the expansion itself. The red shift observed from supernovae in far galaxies shows that, in contrast with Friedman's prediction of a slowing expansion, this is not only not slowing but is accelerating! It is 'as if' an unknown 'pressure' (that is energy) existed inside the universe pushing the galaxies apart more efficiently than their gravitational attraction, acting as the 'engine' behind the expansion. This unknown 'thing' has been called Dark Energy and recent estimates show that it is the main constituent of the universe (68.3%), followed by another unknown constituent, Dark Matter (26.8%), while our good, old, 'ordinary' matter only accounts for (4.9%) of the universe.

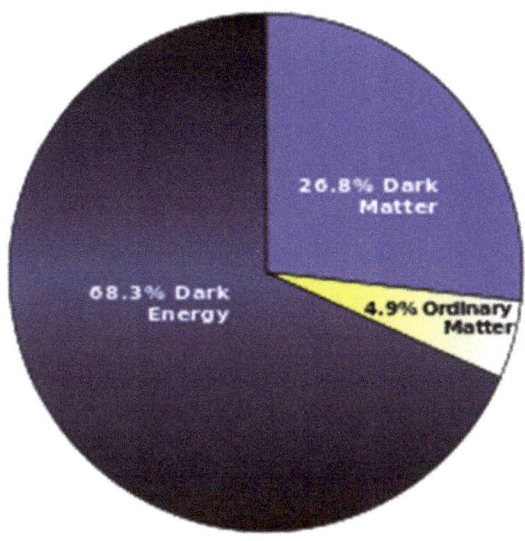

An estimate (2013) of the mass-energy distribution in the universe.

The need for the existence of Dark Matter also arises from the study of galaxies, which has shown that galaxies must contain much more matter than what we can see, matter not directly observable ('dark') since, contrary to ordinary matter it neither emits nor absorbs radiation, only gravitational effects reveal its presence in the observations, in particular these show that:

– if galaxies were made only of visible matter, their mutual gravitational attraction would not be sufficient to explain the stability of clusters of galaxies.

– the amount of visible matter, formed by baryons, is not sufficient to provide the gravitational attraction needed for galaxies to maintain their integrity instead of diffusing into space.

– visible matter is not sufficient to explain the formation of galaxies and clusters of galaxies in the time span from the Big Bang. Their existence requires either more matter than what we see or more time, that is the universe should be older than presently assumed, in contrast with other observations.

– the (almost circular) orbits of the most external stars in elliptical galaxies are determined by the amount of matter inside the whole galaxy and the orbits which have been observed require much more matter inside the galaxies than visible matter.

A further proof of the need for dark matter comes from the observations of gravitational lensing, due to the bending of light rays in a gravitational field as required by general relativity. When the light from a far galaxy passes in the gravitational field of a galaxy nearer to us it is bended, as a lens would do, and from the amount of bending it is possible to calculate the intensity of the gravitational field, that is the mass of the galaxy acting as a lens. The observation of thousands of images of gravitational lensing has confirmed that the visible matter is never sufficient to explain the amount of light bending, a large amount of dark matter is needed to reach the galactic masses consistent with the observations.

In conclusion, how much do we know today about the universe? Quite a lot if we ignore these 'dark things', including a theory which explains with great success how the universe evolved from the Big Bang to become what we observe today.

Not much if we consider that just giving names to these unknown things, Dark Matter and Dark Energy, does not solve the problem.

6. Extra-Terrestrial Life: The Solar System

Innumerable suns exist, innumerable earths orbit around these suns, like the seven planets orbit around our sun. Living beings inhabit these worlds [Giordano Bruno: De Infinito, Universo e Mondi]

A myriad of bodies and corpuscles orbit around a main sequence star classified as a yellow dwarf, the sun, located in the Orion arm of a spiral galaxy known as the Milky Way, at an average distance of about 26,000 light years from its centre and making a complete revolution every 230 million years, with an average revolution speed of about 250 km/s.

The largest bodies in this 'solar system' are the planets which, with reference to their distance from the sun, are called internal: Mercury (the closest to the sun), Venus, Earth, Mars, and external: the giant Jupiter, Saturn, Uranus, and Neptune.

Then there are five 'dwarf planets': Ceres in the asteroid belt and four externals to Neptune's orbit, Haumea, Makemake, Eris and Pluto, which was classified as the ninth planet before 2006.

The Solar System

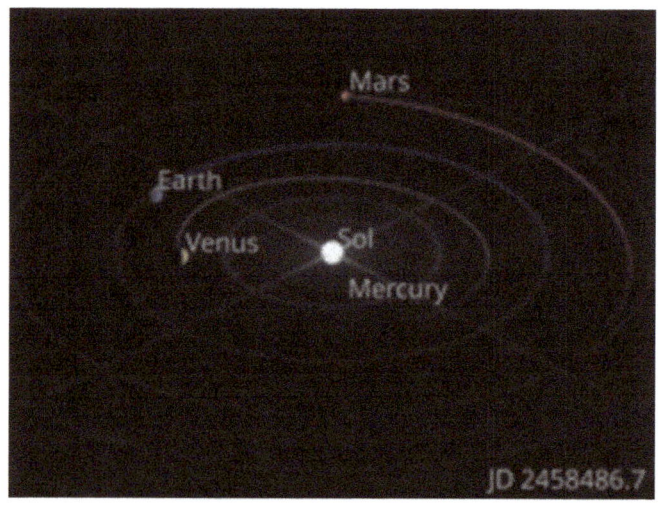

The orbits of the internal planets.

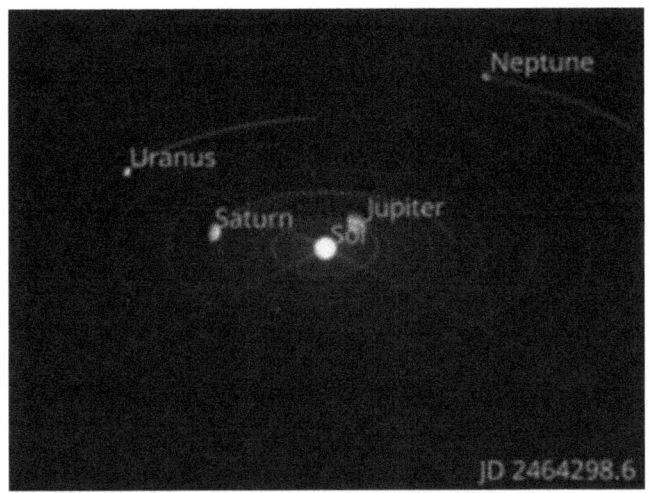

The orbits of the external planets

And finally, an enormous number, though still unknown, of 'minor' celestial bodies, comets and asteroids. They are all gravitationally bound to the sun, and the planets, while travelling along their orbits around the sun, rotate around their axis.

The internal planets and Pluto are solid, with a well-defined surface, and are therefore called 'rocky planets', while the larger planets are fluid, with a gaseous external layer and no definite surface.

The outer planets (in the background) Jupiter, Saturn, Uranus and Neptune, compared to the inner planets Earth, Venus, Mars and Mercury (in the foreground).

Many planets have smaller rocky bodies orbiting around them, the satellites, and all the external planets are surrounded by rings, formed by fragments of rock and ice, constituting a 'belt' rotating all together around the planet. The most famous rings are those of Saturn. The earth has one satellite, the moon, Mars two, Deimos and Phobos, Mercury and Venus have none.

The external planets have a large number of satellites, 69 Jupiter (Ganymede, Callysto, Io, Europa are the largest), 62 Saturn (Titan, Enceladus), 27 Uranus (Titania, Oberon, Ariel, Miranda), 13 Neptune (Triton). Pluto and Charon seem to form a double system, which is surrounded by four satellites (Styx, Nix, Kerberos, Hydra).

Pluto and Charon in a photo from the New Horizon probe

Voyager, the NASA space probe launched in 1977, travelled through the whole solar system visiting also many satellites, and from the images it sent, it was understood that contrary to our moon, some satellites, like Miranda and Triton, still show a substantial volcanic activity.

Saturn with rings

Between the orbits of Mars and Jupiter, we find the 'main asteroid belt', small rocky fragments with dimensions from a few centimetres to hundreds of kilometres, probably the residues of a planet which did not form. The largest asteroid is Ceres, with a radius of 1,000 km, classified today as a dwarf planet. Other groups of asteroids, the Jupiter 'Trojans' are found in more external orbits, separated into two groups, Trojans and Greeks, with their names deriving from the Trojan or Greek heroes of the Iliad, Achilles was the first to be discovered in 1906.

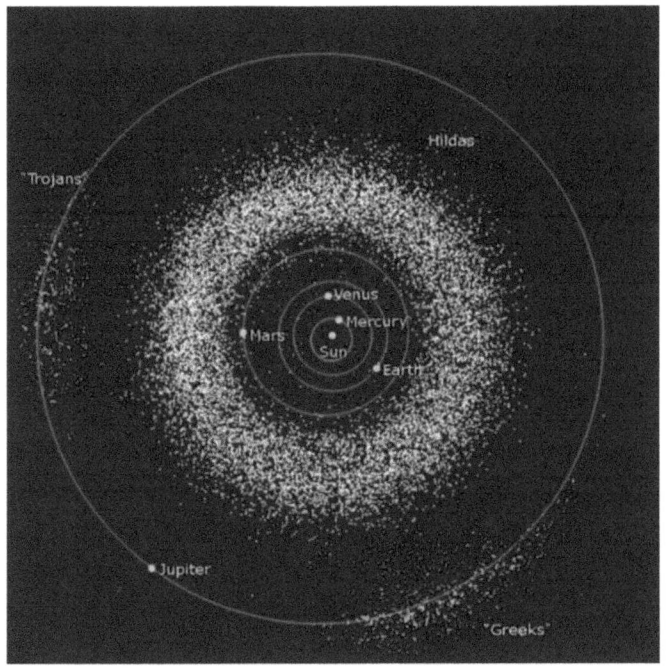

The asteroid main belt and the Trojan and Greek belts

The density of bodies in the asteroid belt is very low, as proven by the fact that our space-probes crossed this region several times with no damage.

Further out, beyond Neptune's orbit, there is another belt, the Kuiper belt, 20 times more extended and about 100 times more massive than the main belt, formed by smaller bodies, mainly ice from ammonia, methane and water. It is also probably the residue of frozen bodies that could not coalesce into a planet.

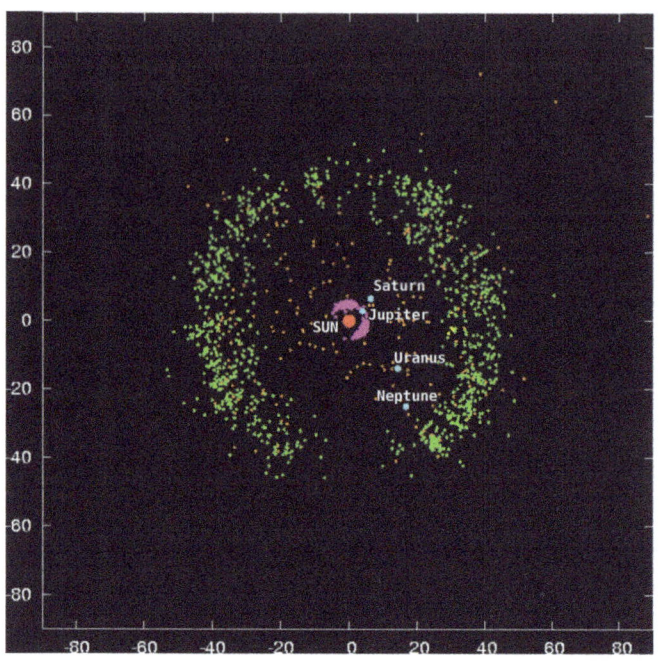

The Kuiper belt.

Still further out, at about one to two light years, the solar system seems to be surrounded by a cloud, called 'Oort's cloud', formed by billions of icy objects, the comets, that occasionally leave the cloud entering the solar system along orbits dictated by the sun's gravity.

The whole system originated from the condensation of a nebula constituted by elements with the typical cosmic abundance (hydrogen, helium and traces of heavier elements) and dust. A region of higher density formed at the centre of the nebula, perhaps due to the shock waves from the explosion of a nearby supernova, and the entire nebula then collapsed on the central region due to its gravitational pull. The temperature increases during the collapse in the central region where, at some point, the temperature needed for the fusion of hydrogen is reached and the sun is born. The rest of the nebula cannot escape from the gravity of the new central star and keeps orbiting around it, forming in time the other bodies of the system, with a series of collision and condensation processes.

Artist view of the primordial solar system

When did all this happen? The oldest meteorites we know have been dated, with radiometric measurements, 4.6 billion years. If all bodies in the solar system, including meteorites and planets, formed in the first stages of condensation of the solar nebula, we can conclude that the system is at least 4.6 billion years old.

How long will the solar system live? For about 5.4 billion years more. Then, as we have seen in the discussion on stellar evolution, the hydrogen in the sun's core will be exhausted and the sun will evolve into a red giant, expanding to swallow the internal planets, while frozen objects further away, like Pluto, could become habitable.

But life on earth will have disappeared long before. In the course of its evolution towards the red giant stage, as the hydrogen in the core keeps decreasing, the dimensions and brightness of the sun will continuously increase, and already in one billion years the increased radiation will make the earth's surface inhabitable; in 3.5 billion years, the oceans will have evaporated, the atmosphere disappeared. Life, at least as we know it, will be impossible.

Life was born and survived on earth, thanks to the solar energy, the presence of oxygen which, forming the ozone layer, has provided the protection against the ultraviolet radiation, and the terrestrial magnetic field which forms a shield against the continuous bombardment from the solar wind.

Could it be born also somewhere else? Consider the solar system first.

For many years, astronomers believed that Mercury rotates on its axis once for each orbit around the sun, always keeping the same face directed towards the sun, in the same way that the same side of the moon always faces earth. This was because, whenever Mercury was best placed for observation, it was always nearly at the same point, hence showing the same face. But radar observations in 1965 proved that the planet rotates on its axis three times for every two revolutions around the sun. That is, an inhabitant of Mercury would only see three days in two years!

Mercury therefore orbits with one hemisphere almost always facing the sun, so that the temperature on this 'day' side is about 350 degrees, while on the 'dark' side, it is close to zero. It has a tenuous atmosphere of carbon dioxide. Too hot for life on the sunny side and photosynthesis cannot take place in the dark side, although it cannot be excluded that some form of microbe can live in these conditions.

Venus is covered by thick, impenetrable clouds of carbon dioxide and water vapour at a temperature of about 40 degrees below zero, and the observations with radio-telescopes have shown that the temperature of the surface under the clouds varies between 200 and 700 degrees, probably due to the 'greenhouse effect' caused by the clouds that do not allow the re-emission of the absorbed radiation. Venus too therefore does not look like a life supporting environment, but a recent study suggested the presence of one potential life sign, a gas called phosphine in the Venusian atmosphere, which on earth is produced by bacteria. Only further investigation will offer a definite answer.

The external planets have no solid surfaces, and their atmospheres are essentially made of hydrogen and helium, with traces of ice crystals from methane, ammonia, perhaps water, at a temperature of about 100 degrees below zero. But it is also possible that deeper inside the temperature could be higher, if so there may be regions where ammonia, methane and perhaps water are fluid, that is areas where today the same conditions are found as the ones of the primordial earth, appropriate for the birth of life. Of course, this is a highly speculative possibility but still it cannot be excluded a-priori that somewhere in the external planets at least the first replicants have been-or will be-formed.

On some of their moons there could be—or have been—the appropriate conditions for the origin of life.

Beneath the frozen surface of Europa, a moon of Jupiter with an icy shell, space probes have detected evidence of a vast ocean of liquid water and it is believed that also two other Jovian moons, Ganymede and Calisto, have subsurface oceans.

Among the moons of Saturn, Enceladus also hides a liquid water ocean beneath an icy shell and jets of ocean water shoot through cracks in the moon's surface, in fact the material from Enceladus's jets forms one of Saturn's rings. NASA's Cassini spacecraft flew through the plume detecting the presence of complex organic molecules, salts similar to those in earth's oceans.

Then there's Titan with an atmosphere dominated by nitrogen, as it does earth's. This moon has a whole hydrologic cycle, with rain, lakes and rivers made of the hydrocarbons, methane and ethane, which are liquid at Titan's temperature of -180 degrees centigrade, but it also possesses a subsurface ocean of water.

Can life develop in liquid hydrocarbons? After all, there is plenty of carbon, fundamental—as we have seen—for the origin of life. Some researchers have argued that life with a completely different metabolism is possible, but at such low temperature chemical reactions proceed at extremely slow pace and the birth of life in this primordial broth of hydrocarbons-albeit completely different from life as we know it could require times much longer than the age of the solar system.

Some of Titan's mysteries will perhaps be solved by the NASA mission 'Dragonfly', a rotary flier that will hop from place to place on Titan's surface.

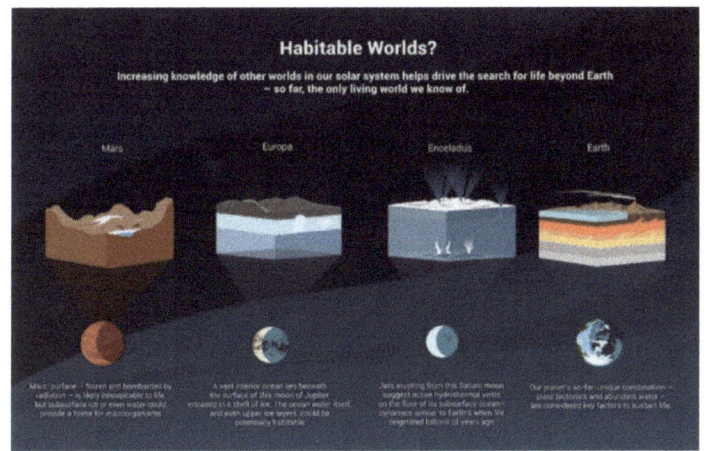

Image credit: NASA/JPL-Caltech/Lizbeth B De La Tor

In conclusion in the solar system, the best candidate for life is Mars. It is even possible that, before appearing on earth, life formed on Mars, since this planet cooled hundreds of million years before the earth and the processes leading to life started therefore when the earth was still too hot.

From the space missions to the red planet, we know that there is water on Mars. The discoveries on the Martian soil of hematite, a mineral which can form only in the presence of water, and of the furrows formed by the erosive action of a liquid, tell us that, at least in the past, liquid water was present on Mars' surface.

The furrows, originating from the crater's edge, have been attributed to liquids (possibly water) running on Mars' surface.

The ESA and NASA space probes that reached Mars have then confirmed that water is still there. Under the planet's crust, at a depth of about 2 km, a frozen lake extending for up to 250 km has been discovered. And the Phoenix probe has revealed ice just two centimetres under the Martian surface. Some photos show that still today water pours out of some slits leaving deposits on the soil.

In conclusion, the basic conditions for the origin of life, including the presence of liquid water, were present both on the earth and on Mars, but since on Mars the chemical evolution had a time of hundreds of million years longer than on earth, the birth of the improbable first replicant is even more probable on Mars than on earth.

The NASA probe-robot 'Curiosity' is walking on the Martian soil collecting rock samples. Their analysis has shown the presence of sulphur, hydrogen, oxygen, carbon and phosphorus, the basic ingredients for the origin of life, and conditions adequate for a microbe's life have been confirmed. We must therefore expect that on Mars there is—or was—some life form which, contrary to what happened on earth, stopped at a lower level. Any further evolution was impossible because the solar wind, which on Mars is not shielded by a magnetic field as on earth, has 'brushed away' the Martian atmosphere, thus making the conditions on the red planet hostile for the development of more complex life forms.

This high-resolution still image is part of a video taken by several cameras as NASA's Perseverance rover touched down on Mars on 18 February 2021. A camera aboard the descent stage captured this shot. *Credits: NASA/JPL-Caltech*

The data from the Curiosity rover suggest that Mars could have supported microbial life billions of years ago. NASA's fifth rover to Mars, Perseverance, successfully touched down on the Red Planet on 18 February 2021, carrying new scientific instruments to build on Curiosity's discoveries.

A primary objective for Perseverance's mission on Mars is searching for evidence that microbial life was present on Mars billions of years ago. It is not possible to bring to Mars the complex instrumentation required to definitively prove microbial life once existed, therefore Perseverance will collect rock core samples to be returned to Earth in future missions for in-depth analysis.

According to mission scientists at NASA/JPL the landing site for Perseverance, the 45-kilometer-wide Jezero Crater, 3.5 billion years ago was the site of a large lake, and even though the water is long gone, evidence that life once existed there could still be found in some feature that couldn't be attributed to anything other than microbial life, a stromatolite for instance. Stromatolites are rocky mounds with typical wavy forms formed long ago by microbial life and are found on earth along ancient shorelines.

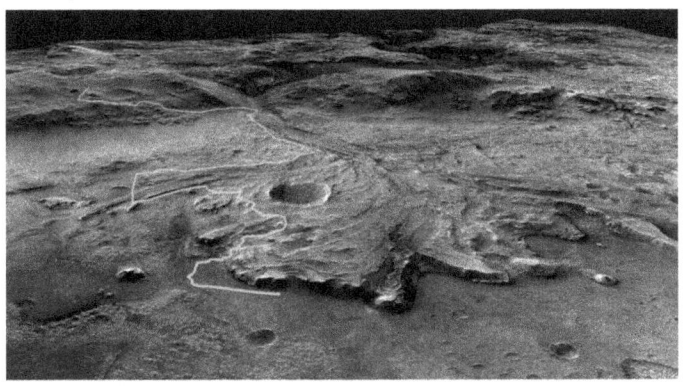

This annotated mosaic depicts a possible route the Mars 2020 Perseverance rover could take across Jezero Crater. *Credits: NASA/JPL-Caltech*

The discovery of stromatolites along the rim of this ancient lake would be a decisive step forward in the search for non-terrestrial life.

If confirmed, the presence of life on Mars, even at the most elementary level as a bacterium, would be the 'proof' that, given the necessary 'initial conditions', the birth of life is the rule rather than an exception.

And this would give extra weight to the possibility that life originated also outside the solar system, as the continuous discovery of exoplanets also seems to suggest.

7. Extra-Terrestrial Life: The Exoplanets

Planets orbiting around stars outside the solar system are called exoplanets.

The first exoplanet, 51 Pegasi b, was discovered in 1995 by the astronomers Michel Mayor and Didier Queloz at the La Silla Observatory in Chile: it is a gaseous giant, but smaller than Jupiter, orbiting around a star similar to our sun, at a distance of 50 light years.

Five gaseous planets orbit around the star 55 Cancri A, a double star in the Cancer (Crab) constellation, at 40.9 light years from the solar system. The first, 55 Cancri b (Galileo), was discovered in 1997, two more, 55 Cancri d (Lippershev) and 55 Cancri c (Brahe) were discovered in 2002. Two years later, 55 Cancri e (Janssen) was discovered, a planet with a mass comparable to Neptune's mass. Finally in 2007, looking back at the data from the Hubble Space telescope, the fifth planet 55 Cancri f (Harriot) was discovered in the habitable zone.

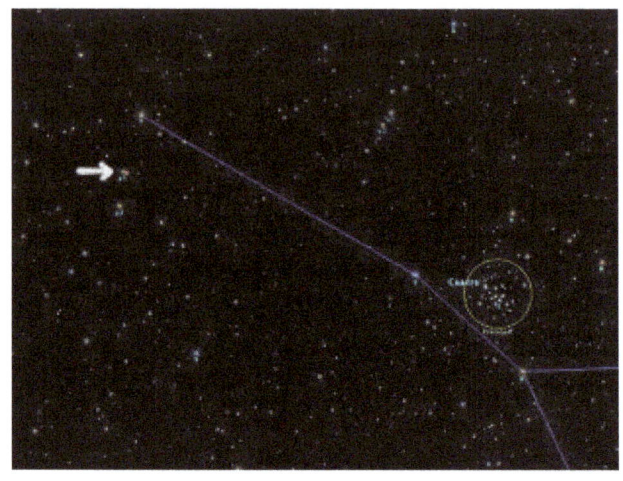

The position of 55 Cancri, in the Crab constellation.

Schematic representation of the 55 Cancri system

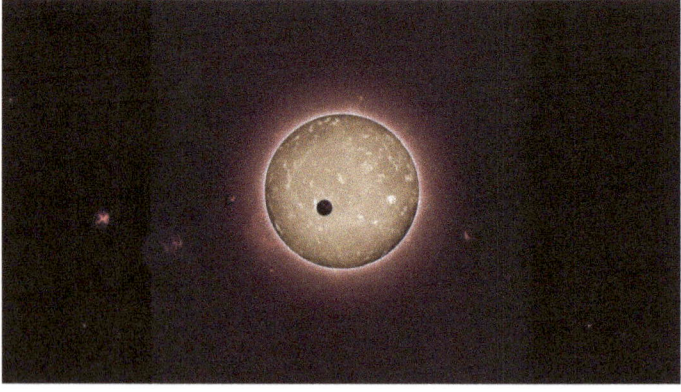

The oldest known system of exoplanets, 55 Cancri, has 5 planets with dimensions like the earth.

In the two decades following the first discovery, more than 3,000 exoplanets have been discovered. Since the region of our galaxy where the observations were made is very small compared to the dimensions of the whole galaxy, it is reasonable to assume that the existence of exoplanets is the rule rather than the exception.

This means that, *in our galaxy alone*, there might be billions of exoplanets. Multiply by the billions of galaxies in the universe…

51 Pegasi b was 'relatively easy' to discover: being a large planet and very close to the central star (it has a period of only four days), in the course of its orbital motion it induces, due to its large gravity, oscillations in the velocity of the central star which are observable even from a ground-based telescope. This method of the 'oscillating star', one of the four methods used today to discover exoplanets, is since the radiation emitted by a moving body 'changes colour' for the observer, that is the receiver (the telescope) sees changes in the wavelength ('colour') of the light from the central star when it oscillates due to the perturbations induced by the motion of the planet.

These oscillations are smaller for smaller planets or for planets far from the central star, since the gravitational effect depends directly on the planet's mass and inversely on the square of its distance from the star, and the observation of the colour change of the stellar light becomes therefore more difficult until it becomes impossible to measure for ground-based telescopes.

But from the discovery of 51 Pegasi b astronomers understood that the presence of giant planets, close enough to their star, could be found in old data on stellar velocities collected by terrestrial telescopes and the 'hunt' for stellar oscillations in the data began: soon tens and then hundreds of exoplanets 'popped up' from the data.

The research based on terrestrial telescopes, mainly by the astronomers Paul Butler and Geoff Marcy, led already in 1996 to the discovery of the exoplanets 70 Virginis, with an orbit of 116 days, and 47 Ursae Majoris, with an orbit of 2.5 years.

In the following decade, hundreds of exoplanets were discovered with observations from ground-based telescopes. By 2017, this method had produced the discovery of 646 exoplanets.

Then the age of space-telescopes began, these have the enormous advantage to observe objects without the filter of the terrestrial atmosphere. With their much larger 'seeing' power than ground-based telescopes, space-telescopes searched for exoplanets via the 'transit method', which involves detecting when the observed light of a parent star dims because of a planet orbiting between the star and Earth.

The age of space-telescopes was opened in 2006 by the European satellite CoRoT, soon followed by the NASA telescopes Kepler, Hubble, and Spitzer.

More than 3000 exoplanets have been discovered to date (2022) and there are still 2000 'suspected' cases waiting for a confirmation.

Apart from the general interest for the study of the exoplanets and the information they can provide about the formation, structure and evolution of 'solar systems', of particular interest is the research of exoplanets on which some form of life might exist.

The only form of life we know is the one on earth and this requires the presence of liquid water. Therefore, leaving aside for the moment the research of 'exotic worlds' (we would not even know what to look for), the research aims to discover exoplanets with the following characteristics:

I. They orbit their star in the 'habitable zone', defined as the region where the temperature is compatible with the presence of liquid water: they should not be too 'close' to their star (water evaporates) or too far (water freezes).

II. The nature of the planetary surface must allow the pooling of water (lakes, seas…): this excludes giant gaseous planets, like 51 Pegasi b, the best candidates are 'small' earth-like planets with solid surfaces (rocky planets).

III. The location of the habitable zone depends on the nature of the central star: it will be farther away from a hot giant star than from a cold dwarf star.

IV. Our sun has been shining for 4.5 billion years and will keep shining for as many years more, then it will evolve into a red giant swallowing part of the solar system. Stars brighter than the sun shine for shorter times, the brightest ones only a few million years, then 'exhaust their fuel' and die. Considering that the appearance of multi-celled organisms on earth required billions of years, on a planet orbiting a star much brighter than the sun, there would be not enough time for life to evolve. May be hundreds of million years are sufficient for the formation of life at the single-cell level, we still do not know when these first forms appeared on earth, but even in this case these primordial forms would be extinct, with the central star, long before complex multi-celled organisms could be formed.

V. The long times needed for the birth of life also require that the central star be 'well behaved' for long times, that is it must be stable for billions of years, radiating its energy at a constant rate. 'Young' stars are unstable, they frequently emit large bursts of energy, called flares, which can alter the conditions on the planetary surface and destroy life, if present.

Our sun also has flares, but the emitted amounts of energy are much smaller than the ones in the flares of young stars, although sufficient sometimes to generate problems here on earth, for instance, to communications.

In conclusion, the following picture for the research of exoplanets that host (might have hosted or will host) life emerges:

Small rocky planets in the habitable zone of stars with an age equal or larger and with dimensions and luminosity equal or smaller than our Sun.

A large part of the exoplanets discovered so far are gaseous or frozen giants, outside the habitable zone of their stars, but some rocky planets have already been discovered.

Already in 2015, from the combined observations of the space-telescope Spitzer and the Italian telescope Galileo, situated in the Canary Islands, the first rocky planet was discovered, HD219134b, at a distance of 'only' 21 light years from us but outside the habitable zone of its star.

Kepler-186f was the first rocky planet to be found within the habitable zone—the region around the host star where the temperature is right for liquid water. This planet is also very close in size to earth.

In 2016, astronomers reported the discovery of a planet orbiting around Proxima Centauri, the closest star to the solar system (four light years). This planet, named Proxima b, has a mass slightly larger than earth, and its orbit is closer to its central star than Mercury's orbit to the sun (its period is just 11 days) but, since Proxima Centauri is a dwarf star cooler than the sun, the conditions on its surface might still allow the presence of liquid water, that is the possibility of life 'in our backyard'.

Artist's image of the surface of the planet Proxima b, orbiting around the red dwarf Proxima Centauri, the closest star to the solar system. Credits: ESO/M Kornmesser

A planetary system with characteristics similar to the solar system has been discovered by Stratospheric Observatory For Infrared Astronomy (SOFIA): the central star is Epsilon Eridani (eps Eri), 10.5 light years away in the constellation of Eridanus.

The presence of a planet, Epsilon Eridani b, with a mass similar to Jupiter's and orbiting around eps Eri at a distance of the order of Jupiter's distance from the sun, has also been reported. SOFIA has also confirmed that around eps Eri there is a ring formed by dust, gas and rocky debris, like the asteroid belt of our solar system, in an orbit between the star and Epsilon Eridani b, as the asteroid belt is found in an orbit between the sun and Jupiter.

Artist's illustration of the Epsilon Eridani system showing Epsilon
Eridani b. In the right foreground, a Jupiter-mass planet is shown
orbiting its parent star at the outside edge of an asteroid
(web NASA.gov)

Epsilon Eridani is a system younger than the solar system
and, given the similarities in structure, observing this system
is like observing a 'young solar system', when the internal
planets had not yet been formed.

In February 2017, NASA announced the discovery of a
system, TRAPPIST-1, formed by seven rocky planets with
dimensions similar to Earth, named TRAPPIST-1b...
TRAPPIST-1h, *three of which* (TRAPPIST-1f, g, h) are in the
habitable zone of their star, a cold dwarf at a distance of about
40 light years from us, with an age about twice the sun's age
and with a mass of about 8% of the solar mass.

All the conditions listed above are met for TRAPPIST-1, whose discovery is due to a particularly efficient collaboration between space-and ground-based telescopes, in this case, the space-telescope Spitzer and the ground-based telescope Transiting Planets and Planetesimals Small Telescope (TRAPPIST), operating in Chile.

Artistic view of the planetary system TRAPPIST-1
(website NASA.gov)

The cold dwarf stars, like TRAPPIST-1, are stable, their temperature and luminosity stay nearly constant for billions of years, allowing life all the time to evolve.

But this also means that the planets of TRAPPIST-1 have absorbed the stellar radiation for billions of years, causing the possible evaporation of any atmosphere or water present on their surface. This is what happened in our solar system to Mars: there was water in the past on its surface, but it evaporated, as did the Martian atmosphere, after billions of years of exposition to the solar radiation.

The present hypothesis is that only on the two far away planets, TRAPPIST-1g and TRAPPIST-1h, liquid water might still be present.

These two planets have a density lower than the earth's density and therefore a lower gravity: this could imply that the most volatile molecules escaped from their surface and formed a dense atmosphere which 'shielded' the planet from stellar radiation, thus allowing water on the surface to stay liquid.

The ultraviolet radiation emitted by the star is mainly responsible for the 'breaking' of water molecules into their constituents, hydrogen and oxygen, and these can escape from the planet's gravity if they receive sufficient energy from the most energetic part of the stellar radiation spectrum, like X-rays.

The amount of ultraviolet radiation emitted by TRAPPIST-1, measured by our telescopes, provides the basis for an estimate of the amount of water possibly lost by the TRAPPIST-1 planets.

In this way, it has been estimated that the most internal planets, TRAPPIST-1b and TRAPPIST-1c, which receive the greatest amount of ultraviolet radiation, might have lost, in the past eight billion years, an amount of water of the order of 20 terrestrial oceans, while the more external planets probably lost much less water and pools of liquid water could still be present on their surfaces.

This illustration shows the possible surface of TRAPPIST-1f, one of the newly discovered planets in the TRAPPIST-1 system. (Website NASA.gov)

For TRAPPIST-1, the system with the largest number of 'earth-like' planets discovered so far, new results are expected from future space-telescopes, like the JAMES WEBB and the Wide-Field Infrared Survey Telescope (WFIRST), which will carry the technology (coronographers, star shade) needed for the direct observation of exoplanets.

In the meantime, Hubble discovered an exoplanet with water in its atmosphere: it is WASP-121b, a member of the system WASP-121 at about 900 light years from us. It is a planet with a mass slightly larger than Jupiter's mass and with a radius almost twice as large as Jupiter's radius, its density is therefore much smaller than Jupiter's density. It is very close to the central star, with a period of just 1.3 days as compared to the 12 years of Jupiter's period around the sun.

The spectroscope on board the Hubble Space telescope could detect water molecules in the external layers of this planet (the stratosphere) and estimate the temperature, which can reach 2500 degrees Celsius, making WASP-121b a 'boiling Jupiter'. Certainly not the right environment for life.

The exploration goes on with the present and future space-telescopes, like Transiting Exoplanet Survey Satellite (TESS), which will proceed to a complete observation of the nearest and brightest stars searching for planets, selecting the most promising for the existence of life, which will then be the object of a more detailed exam from the WEBB space-telescope, equipped with a mirror so advanced to be able to capture directly the light emitted by a planet.

The present and future NASA missions for the study of exoplanets.

In the meantime, TESS, designed and launched to discover, via the transit method, planets in the habitable zone of nearby stars, has made several important discoveries.

In 2018, a rocky, earth-like planet named LHS 3844b, was discovered around an M-dwarf star 48.6 light years away. According to follow-up observations by the Spitzer space telescope, this planet has little, or no atmosphere and its surface is probably similar to our moon's surface, that is covered in the same cooled volcanic material. M-dwarf stars are the most common type of stars in our galaxy and may therefore host a large number of the exoplanets in the Milky Way, but these stars emit an intense ultraviolet radiation which can brush away their atmospheres and is dangerous for life, at least for life as we know it.

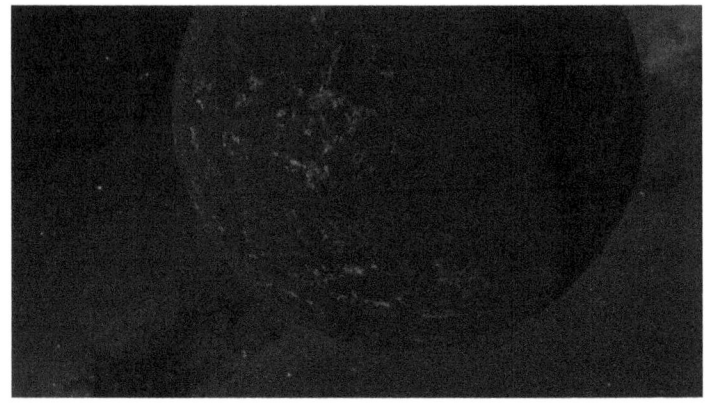

Artist's view of the exoplanet LHS 3844b
Credits: NASA/JPL-Caltech/R. Hurt (IPAC)

In 2019, TESS discovered a new planet, called L98-59b, with a size between the sizes of Mars and earth, in a system of three planets (L98-58c, L98-59d) orbiting around a nearby bright, cool star L98-59, again an M-dwarf 35 light years away.

The planets in the L98-59 system compared to Mars and earth in order of increasing size. *Credits: NASA's Goddard Space Flight Centre*

This planet is the tiniest discovered by TESS so far (80% Earth's size), while the other two planets in the system, c and d, are 1.4- and 1.6-times earth's size respectively. None of them lies within the star's habitable zone.

Data collected by TESS also allowed to discover the first planet orbiting two stars (one larger and one smaller than the sun) in a system 1300 light-years away in the constellation Pictor. It is called TOI 1338b, with a size between Neptune and Saturn (about seven times larger than earth's). The two stars form what is called an 'eclipsing binary system', that is, we see an eclipse when they circle each other in our plane of view.

The circumbinary planet TOI 1338 b is orbiting two stars *Credits: NASA's Goddard Space Flight Centre*

The most exciting discovery by TESS so far is probably its first earth-size planet in the habitable zone of its star. It is called TOI 700d and joins the planets in the Trappist-1 system as one of the few earth-size planets in the habitable zone discovered so far. The system is formed by three planets (b, c, d) orbiting the small, cool M-dwarf star TOI 700, about 100 light-years away in the constellation Dorado.

After the discovery by TESS, the Spitzer space telescope allowed to establish that one of the three planets, TOI 700d, is in the habitable zone. The innermost one, TOI 700b, is almost exactly earth-size, probably a rocky planet, while the middle one, TOI 700c, is about 2.6 times earth's size and probably gaseous. Only the outermost one, TOI 700d, about 20% larger than earth, is in the habitable zone of TOI 700. It also receives from its star about 90% of the energy that the earth receives from the sun. These look like ideal conditions for life, but the problem is that their 'small' orbital periods (10, 16 and 27 days respectively) suggest that they may all be tidally locked to their star, meaning that one side is constantly in daylight, while it is constantly 'night' on the opposite side. If so, one side of TOI 700d may be too hot, or receive too much ultraviolet radiation, for life and the other side may be too cool for life. We will know…tomorrow.

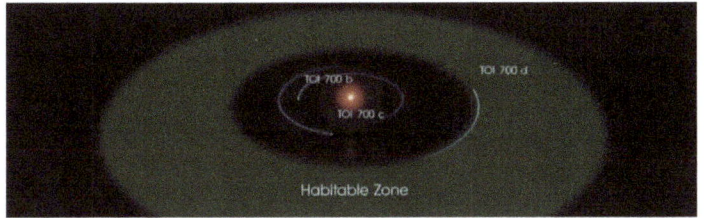

The TOI 700 system

The latest discovery by TESS (2021) is a planet about half the size of Saturn in a triple-star system: KOI-5Ab. When the signal from this planet was first observed by the Kepler space-telescope in 2009, it appeared to be circling one of the three stars, but Kepler's data were not sufficient to determine whether the signal was actually an erroneous glitch from one of the two other stars, or, if the planet was real, which of the stars it orbited.

Thanks to new observations from TESS, combined with data from ground-based telescopes, the WM Keck Observatory in Hawaii, Caltech's Palomar Observatory near San Diego, and Gemini North in Hawaii, David Ciardi, chief scientist of NASA's Exoplanet Science Institute and co-workers were finally able to untangle all the evidence surrounding KOI-5Ab and prove its existence. As Kepler had observed previously, TESS found that the planet orbited its star roughly every five days.

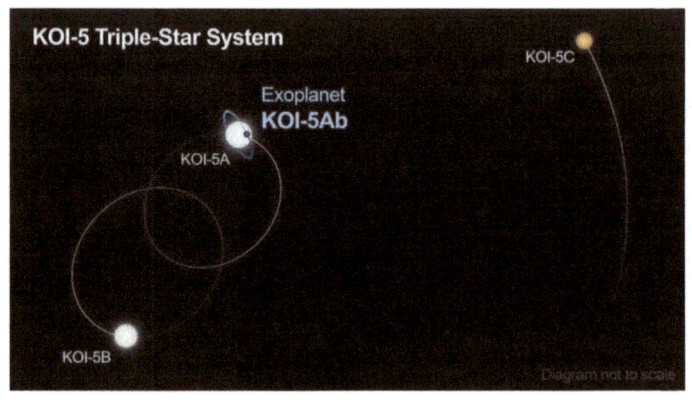

The KOI-5-star system consists of three stars, labelled A, B, and C in this diagram. Star A and B orbit each other every 30 years. Star C orbits stars A and B every 400 years. The system hosts one known planet, called KOI-5Ab, which was discovered and characterised using data from NASA's Kepler and TESS missions, as well as ground-based telescopes. KOI-5Ab is about half the mass of Saturn and orbits star A roughly every five days. Credit: Caltech/R Hurt (IPAC)

Let us summarise the present situation with exoplanets.

Kepler ran out of fuel in 2018, after nine years of observations which revealed more than 2,800 confirmed planets outside our solar system. Kepler's field of view covered only 0.25% of the sky, which allowed scientists to extrapolate that there are billions of planets in our galaxy—more planets than stars.

The search continued with other space-telescopes and to date (January 2023) the number of confirmed exoplanets is 4331.

These fall into four broad categories: large hot gas giants, like Jupiter, mysterious 'Super Earths', frozen planets, like

Neptune and small, rocky planets in earth's size range (see figure).

The several thousand planets so far confirmed to be in orbit around other stars—exoplanets—fall into four broad categories: large, gas giants, Neptune-like worlds, 'Super-Earths' bigger than earth but smaller than Neptune, and terrestrial planets in earth's size range. Within these categories, however, scientists find even more variety. Among the gas giants, for instance, are 'hot Jupiters', infernal worlds with tight, star-hugging orbits. Credit to Pat Brennan of NASA's Exoplanet Exploration Program.

What can we say about habitability and possible life on this wide range of different planets orbiting stars with a wide range of properties impacting whether the rocky planets in their orbit can support liquid water? What is the occurrence rate of potentially habitable worlds?

According to a recent work (2021) by Steve Bryson and co-workers at NASA, combining Kepler's final dataset with data about each star's energy output from an extensive trove of data from the European Space Agency's Gaia mission, about half the stars similar in age and temperature to our sun could have a rocky planet capable of supporting liquid water on its surface. Their new analysis accounts for the relationship between the star's temperature and the kinds of light given off by the star and absorbed by the planet, and the role of the planet's atmosphere to determine how much light is needed to allow liquid water on a planet's surface and potentially life.

Our galaxy holds an estimated 300 million of these potentially habitable worlds, some are close neighbours, with four potentially within 30 light years of our sun and the closest likely to be about 20 light years from us.

We have searched so far, a minimal part of one of the billions of galaxies in the universe—our Milky Way—and in a couple of decades discovered thousands of exoplanets, some of them in the habitable zone of their host stars. It is then reasonable to assume that there is a very high probability—I would say certainty—that planets like our earth exist in the universe and that on some of them some 'form of life' was (or will be) born and evolved.

But these 'forms of life' have certainly no resemblance to life as we know it because, as discussed before, the probability of the appearance of homo sapiens, even starting from the same 'initial conditions' of the primordial earth, is vanishingly small, practically null. Homo sapiens is alone in a universe teeming with life, most probably at bacterial level.

Now perhaps it is possible to give an answer to the question Enrico Fermi posed back in 1950 during lunch with fellow physicists: 'where is everybody'? He argued that even at a slower-than-light pace our galaxy could be easily crossed by a spacefaring civilisation within a few million years and in the 14 billion years since its existence, it should have been crisscrossed millions of times by spaceships or signals from other civilisations.

A possible answer could be: "'everybody is out there indeed," but bacteria do not build spaceships or radar.

Bibliography

Irving Adler (1957) *How Life Began,* J. Day Co.

Isaac Asimov (1984) *Asimov's New Guide to Science (vol. II)* Basic Books Inc.

John D. Barrow (1994) *The Origin of the Universe*, Basic Books Inc.

Bernard Cohen (1974) *The Birth of a New Physics,* Doubleday & Co. Inc.

Charles Darwin (1985) *The Origin of Species,* Penguin Books.

Charles Darwin (1989) *Descent of Man, and Selection in Relation to Sex,* New York University Press.

Charles Darwin (2007) *A Naturalist Voyage... (The Voyage of the Beagle),* London, J. Murray.

Paul Davies (ed.) (1989) *The New Physics*, Cambridge University Press.

Richard Dawkins (1991) *The Blind Watchmaker,* Penguin Books.

Jared Diamond (1998) *Guns, Germs and Steel,* Vintage.

Jared Diamond (1992) *The Rise and Fall of the Third Chimpanzee,* Vintage.

J.L.E. Dreyer (1906) *History of the Planetary Systems from Thales to Kepler,* Cambridge University Press.

Stephen Jay Gould (2000) *The Lying Stones of Marrakesh,* Jonathan Cape.

Stephen Jay Gould (1992) *Bully for Brontosaurus,* Penguin Books.

Stephen Jay Gould *Eight Little Piggies,* Penguin Books 1994.

Stephen Jay Gould (1990) *An Urchin in the Storm,* Penguin Books.

Stephen Jay Gould (1989) *Wonderful Life. The Burgess Shale and the Nature of History,* New York, Norton & Co.

Alexander Koyrè (1957) *From the Closed World to the Infinite Universe,* J. Hopkins Press.

Thomas S. Kuhn (1957) *The Copernican Revolution. Planetary Astronomy in the Development of Western Thought*, Harvard university Press.

Jacques Monod (1971) *Chance and Necessity: An Essay on the Natural Philosophy of Modern Biology,* New York, Knopf.

Slkovskii, S., Sagan, C. (1966) *Intelligent Life in the Universe,* New York, Dell Pub. Co.

Arnold J. Toynbee (1976) *Mankind and the Mother Earth,* Oxford University Press.

Steven Weinberg (1993) *The First Three Minutes,* Basic Books.

Figures in the text are from Wikipedia or from the NASA website.